THE NAKED SAVAGE

Sinclair

5-00

THE NAKED SAVAGE

THE NAKED SAVAGE

ANDREW SINCLAIR

SINCLAIR-STEVENSON

First published in Great Britain by
Sinclair-Stevenson Limited
7/8 Kendrick Mews
London SW7 3HG, England

Copyright © 1991 by Andrew Sinclair

All rights reserved. Without limiting the rights under copyright reserved, no part of this publication may be reproduced, stored in or introduced into a retrieval system or transmitted, in any form or by any means (electronic, mechanical, photocopying, recording or otherwise), without the prior written permission of both the copyright owner and the above publisher of this book.

The right of Andrew Sinclair to be identified as author of this work has been asserted by him in accordance with the Copyright, Designs and Patents Act 1988.

British Library Cataloguing in Publication Data
A CIP catalogue record for this book is available from the British Library.
ISBN: 1 85619 088 9

Typeset by Butler & Tanner Ltd
Printed and bound in Great Britain by
Clays Ltd

Ere the base laws of servitude began,
When wild in woods the noble savage ran.

> from 'The Conquest of Granada'
> by JOHN DRYDEN

Time held me green and dying
Though I sang in my chains like the sea.

> from 'Fern Hill'
> by DYLAN THOMAS

CONTENTS

Prologue
ix

1 WHEN WILD IN WOODS
1

2 THE AMERICAN SAVAGE
28

3 THE SAVAGE WITHIN
47

4 THE SAVAGE ENSLAVED
58

5 THE MYTH OF THE SAVAGE
75

6 THE WILD VERSUS THE MACHINE
106

7 THE GREAT ENGINEER
121

8 THE CULT OF THE SAVAGE
130

9 THE RESPECTABLE SAVAGE
145

10 GREEN AND DYING
162

Notes
174

Index
199

PROLOGUE

'GREEN AS BEGINNING,' Dylan Thomas once began a poem, but of his own ending, he wrote, 'Green and dying.' Yes, we were green at our beginning, when we were savages in the woods. A savage is somebody who lives in the trees, a person who feels a part of all growing things and living things, even though he may gather or hunt them.

Civilisation and pollution followed, and we thought we had contaminated the earth. In fact, we cannot do that. Even human ingenuity may not go that far. The Earth or Gaia will survive, whatever we may do to it. It will replace us with other forms of life, if we manage to destroy our species on this planet. We are not as important as we think we are.

This book tells the story of how we came out of the woods as savages when we were sure that we were only a twig on the great tree of being. We forgot that common feeling with nature during the past brief five thousand years of human civilisation, irrelevant to the long life of other organisms upon this Earth. Now we are again conscious of that old bonding with the trees that breathe and the animals which are also alive. We are they once more, while they have always been like us. We are still savages, even if we exist in cities. This is the hope and the secret for all living things on this planet here and now. We are green as beginning again, if we remember how we were. Otherwise, we are green and dying.

1
WHEN WILD IN WOODS

AS APE-LIKE MEN, we once came out of the caves and the woods, and our unconscious still links us to our origins. Yet there are no records of those times, no evidence of history nor archaeology. Aristotle is the first major thinker to deal with the problem of the superior and inferior man as well as to link our nature to biology. Tacitus is the first historian to speak of the forest culture of the Germanic peoples, Frazer the first social anthropologist to trace the cult of the regeneration of life through tree-worship. Of the earliest life of the forest dwellers, we have only the incomplete evidence of the past ten thousand years, hardly written, occasionally excavated by the new archaeologists.

Yet certain processes took place. After the last Ice Age trees or bush covered most of the fertile land. Primitive peoples who wished to cultivate the earth had to clear it. The residual forest tribes today use metal and fire to cut down the trees and liberate the soil. This slash-and-burn technique provides a fertiliser of ash for the brief fields, which are allowed to revert to woodland again after a harvest. Without the axe and the flame, no large clearing of trees is possible.

The first fully human hunters and gatherers had come out of Africa more than a million years before. But their growth was limited by small parasites, which attacked their bodies and competed for food and energy.[1] When human settlement spread to the northern and temperate zones, naked man had to clothe himself with the skins of wild beasts. He left most of the lethal tropical parasites behind him and flourished, hunting the herds of mammoths and bison on the frozen tundra of the plains. Art began in the caves of the hunters thirty thousand years ago, the flowing finger-marks pressed into damp clay making the face of a musk-ox in the galleries of Altamira in Spain. Rock painting and carving covered the globe wherever hunting societies moved. It displayed the confusion between the human predator and his prey, the worship and feeling of man for beast.

For the hunter senses himself in the process of nature. He is a part of the world he knows. His art indicates that he makes no distinction between himself and the animals, which kill him, or which he kills. Indeed, the paintings show men transformed into beasts and birds, and these into men. The mythology of hunting societies has transference and talk between the animal kingdom and human beings. Early religion and some surviving creeds believe in the migration of souls and spirits from insect and plant and tree, from fish and fur and fowl to mankind. The teaching and art of Taoism in China and of Shintoism in Japan gave human beings a place in nature, not dominion over it. They might be hunters and killers, but only among the other carnivores of the whole of creation.[2]

When these primal human societies began to predominate over other species, they felt dependent upon their prey. The hunter killed in order to live; but he asked the forgiveness of his victims. In his extraordinary works on primitive culture, Sir Edward Burnett Tylor gave examples of hunters world-wide, begging the understanding of their suppliers of meat and hide.

> "Savages talk quite seriously to beasts alive or dead as they would to men alive or dead, offer them homage, ask pardon when it is their painful duty to hunt and kill them. A North American Indian will reason with a horse as if rational. Some will spare the rattlesnake, fearing the vengeance of its spirit if slain ... The Stiens of Cambodia ask pardon of the beast they have killed; the Ainus of Yesso kill the bear, offer obeisance and salutation to him, and cut up his carcase ... As for believers, savage or civilised, in the great doctrine of metempsychosis, these not only consider that an animal may have a soul, but that this soul may have inhabited a human being, and thus the creature may be in fact their own ancestor or once familiar friend."[3]

The ending of the last Ice Age and the change of climate led to the metamorphosis of Mediterranean and European societies. The frozen grasslands were largely replaced by forests, while in the Near East and Africa, the deserts spread their sands over swamps and fertile soil. The great herds of wild beasts shrank. Men and women had to turn to managing their sources of meat and developing the produce of the earth. In Europe and parts of Asia, horses and pigs, sheep and goats, dogs and cats, water buffalo and reindeer, geese and ducks, chickens and even bees were brought under control. In the Americas, however, only llamas and guineapigs, dogs and some fowl were domesticated, while Africa and Australasia still relied on game. Equally, while the river civilisations of China and the Near East used irrigation and the plough to create harvests of rice and wheat, barley and lentils and peas, the American maize and potato were only developed slowly from their wild state. Changed in the main from hunters into foresters and herders and farmers during the ten millennia before Christ, the peoples of the Old World began to outstrip the tribes of the New World. Their previous beliefs in the holiness of all living things, plant and tree and bird and beast, eroded with their massing in towns and cities, and with their dividing from the other creatures which lived in caves and lairs.

When prehistoric men and women were compelled by the climate to move from hunting to agriculture, their first terrible work was to cut the timber for their shelters and to grub the roots from their fields. For hauling out the stumps, they needed metal and flint and domestic beasts. In southern Europe, the charcoal burner and the goat would quickly strip the mountains of timber; but in the north, outside the coastal villages subsisting off the fruits of the sea, thousands of years of toil by men and beasts in the ages of bronze and iron were necessary for the slow growth of villages surrounded by their few fields, small open patches in the blanket of trees that covered all the land between the Vistula and the Rhine. This unceasing labour, allied to the fear of attack from foreign tribes, led to the institution of slavery, by which prisoners of war were set to do the work of domestic beasts while their owners concentrated on defending the tribe. Meanwhile, the first enemy remained the intractable and antagonistic forest.

Early village cultures are now being excavated in the Balkans and Turkey. These date back to the seventh millennium before Christ, and before that. Community recedes further into antiquity. Recent diggings show little groupings of people struggling for survival in a hostile environment. Where the land was already clear, their task was less difficult. But when there was forest and the villagers had to fight it back, the trees of their origins were the frontier of their fears. The savage stood at the edge of the encircling woods. The defences of cabin and clearing and custom were infiltrated by the dark memories of the surrounding wilderness. As the recent villager prayed to the new gods of the hearth, the ancient devils beckoned from the branches. As he bent to the hoe, his dagger pricked at his belt. For the savage stood near the barbarian pioneers on the fringes of civilisation. Reversion to life in the woods was the only refuge from loss in war and the burning of the village. If the newly cultivated trees, which bore fruit, were destroyed, the wild trees were the final shelter. Advance to the city was a long process of accumulating stores of food, free from attack. In that halfway stage between forest and market-place, the barbarian stood, unrecorded, necessary, expectant.

Yet as the new archaeology has shown, the barbaric cultures of prehistoric Europe were capable of architectural and astronomical feats that surpassed those of the first urban civilisations of the Near East. Radio-carbon dating, corrected by calibrations made from the tree-rings of the bristlecone pine, has shown that the temples of Malta, the megaliths of Brittany and the observatory of Stonehenge apparently predate the Pyramids and Mycenae. The Copper Age appears in Balkan villages before Mesopotamian cities. Barbaric societies seem to have been as technologically inventive as urban ones. The tribal organisations of timber and early farming cultures were as capable of storing a food surplus in a chieftain's house and of constructing a ritual centre as any Babylon or Nineveh. Even the one distinguishing element between the early European barbarian and the near Eastern citizen, the use of an alphabet, has been put in question by the discovery of sign writing in the ancient Vinca culture of Romania and Bulgaria. The assumption of the first archaeologists that all Western civilisation was diffused from the Near East is proven untrue.[4] The culture of timber and village in prehistoric Europe was as vital and creative as that of the delta city. The savage and the barbarian contributed as much as urban civilisation in the crawl of men from the cave and the woods to the field and the roof.

Fortunately, the myth of the conflict between the savage and the city is set down in the legends of the earliest urban civilisation of the Near East. In the third millennium before Christ, only one city culture existed outside India and China, that of Mesopotamia. Like China, it depended on the control of water power, and like China, it held itself to be the centre of the earth. A contemporary map showed Babylon as the axis of the world, surrounded by neighbouring cities and the irrigation canals that fed it. Then came the encircling ocean with triangles beyond, representing unknown lands beyond the pale of civilisation, the lands of the inferior and the beast.

The cities of Sumeria in Mesopotamia and later of Egypt were theocracies in which rulers and priests – sometimes combined as in the divine person of Pharaoh – reigned over river cultures by controlling water power and by divining the heavens in order to assure harvests on earth through irrigation. The Sumerian temples were the prototypes for a religion exalting the dominant god-king. The innumerable clay tablets written in cuneiform that recorded the accounts of the state were the necessary method of supporting the government through taxation. They also recorded the amount of livestock and the size of the gathered crops.

WHEN WILD IN WOODS

As the Sumerian poet wrote of the original divine ruler, the creator of agriculture as well as of writing and geometry, and cities and temples:

> The plough and the yoke he directed
> The great prince Enki
> Opened the holy furrows
> Made grain grow in the perennial field.[5]

Within the Sumerian cities, society was divided into free men and slaves. The slaves were not all barbarians; they could be prisoners from the next city or the children of slaves or guarantors for a debt. They were treated as mere units of labour, to be bought, bartered and sold. Although they could trade and even own slaves, they were brutally punished if they tried to escape. Female slaves had to become concubines or prostitutes on request. By this contempt for the lives of fellow human beings, the Sumerians created within their original culture a fatal distinction between citizens and other people. Although slavery may have seemed necessary to provide a labour force to build the cities and to maintain the irrigation works, yet the custom of treating humans as beasts of burden was the first and worst mistake of the city, compounded by the legal division of mankind into citizens and brutes.[6]

This same urban culture also produced the *Epic of Gilgamesh*, among the most ancient of the written myths and the first to deal with the attitude of civilised men, not to slaves but to savages and trees and wild beings. The popularity of the epic was partially due to the fear felt by these ancient wealthy cities for the nomads of the Arabian deserts and for the mountain tribes to the north. Gilgamesh was traditionally the great builder of the walls and temples of Uruk, also the legendary venturer back into the origins of his people. In myth, he went into the depths of the dark forest to fight with evil and bring back cedarwood for the temples. So he united in his royal person both the ancestral hero and the forgotten conflicts of primitive life.

The epic begins with Gilgamesh's encounter with a wild man, Enkidu, who is innocent of humankind and knows nothing of cultivation. He lives with the beasts and his body is covered with matted hair. He is in total communion with nature. On Gilgamesh's advice, a harlot is sent to the wild man. Enkidu stays six days and nights with her, then goes back to the beasts, which now reject him. He returns to the woman, who counsels him to shave and wear clothing and be a shepherd. He gives up sucking the milk of wild animals, and he learns to eat bread and drink wine. Then he challenges Gilgamesh, who wrestles with him in the city and defeats him and becomes his loved companion.

So the epic starts with the myth of the founding of the cities of Mesopotamia, the initial period of savagery, the living like a beast, the slow process of domestication, the learning to eat the foods of the farm, the coming to the city, the challenge to authority and the final acceptance of the power of the king. The epic continues with Gilgamesh deciding to destroy the source of all evil and brutishness, Humbaba, the giant of the distant cedar forest. Enkidu counsels his friend and master against such folly, for he knows the terrible wood, which stretches ten thousand leagues in every direction. Humbaba himself is like fire, like a storm, like the very jaws of death. Yet Gilgamesh is full of Mesopotamian melancholy. In the city, he says, man dies with despair in his heart. He must go out and kill the source of all human evil and cut down the cedar trees for the rebuilding of the city.

At the entrance to the forest, where civilisation ends and savagery begins, the two companions discover a perfect gate of cedarwood. Half-brute, half-man, Enkidu cannot find it in himself to destroy the gate, for he is caught at the point where the wild meets the tame. As he thrusts the gate open, his hand loses its strength; but still he and Gilgamesh go on to battle with Humbaba. Gilgamesh fells Humbaba's first cedar with his axe and the giant comes out to fight. Enkidu, the child of the forest, is terrified; but Gilgamesh encourages him. 'Take your axe in your hand and attack. He who leaves the fight unfinished is not at peace.' The fight itself equates the giant with the forest, evil with the wood. As Gilgamesh cuts a cedar, Humbaba blows out fire to burn him. When Gilgamesh has felled the seventh cedar and has cut and bound all the branches, then Humbaba is delivered to him 'like a noble wild bull roped on the mountain, a warrior whose elbows are bound together.' He begs mercy of Gilgamesh, saying that he is a force of nature like Enkidu, that he is chosen by the gods to be the keeper of the dark forest. If he is freed, he will become Gilgamesh's servant, he will cut down the forest and build a palace from the cedarwood. But Enkidu is jealous of this second tamed elemental power, and so he and Gilgamesh hack Humbaba apart with their axes, and the eighth cedar is dashed to pieces. Instead of containing evil, the companions have loosed it with the giant's death and have enraged the gods. Although their victory allows them to fell the forest and clear the roots as far as the Euphrates – thus creating the fields and building materials for Uruk – yet the wrath of the gods at the death of Humbaba makes them give the giant's fire to the barbarians, to the lion, to the wilderness, to the daughter of death.

Gilgamesh is not only the root epic of all, but the primary legend of man's slow change from a forest hunter to an urban dweller. When the king of the city and the tamed natural man go back to the savagery of the forest in order to battle with the problems of evil, they find that evil exists in the form of life itself. Symbolically, when they cut and bind the wood, they bind the ferocity of those bestial creatures which live in the forest. But when they are offered the psychological choice between the containment of brute instinct or its repression and destruction, they choose the second force, since Enkidu still fears the wildness remaining in him. Thus he and Gilgamesh are able to deny the savagery within themselves, to clear the trees, to make fields as far as the river, to rebuild the city. But the price is the menace of the barbarians and of the wilderness and of death, which will always threaten a soft and civilised people without enough fire in their spirits to resist a fierce attack. The *Epic of Gilgamesh* is a more satisfactory legend of the growth of civilisation than the simplicity of the Garden of Eden and the Fall from grace into hard labour.

Later in the epic, Enkidu dies through disease and is punished for the destruction of the forest, while Gilgamesh mourns him by becoming a wild beast. He lets his hair grow, he wanders through the wilderness in the skin of a lion, he eats the raw flesh of animals. He joins the natural world which he has tried to deny. He even reaches the underworld in search of Enkidu, only to discover that death is the fate of all men. He returns to Uruk to die himself.[7] He has created civilisation from savagery, reverted to the animal and then returned to his culture because of the wisdom of a philosophy that can reconcile him with the impermanence of life. Like the Nebuchadnezzar of the Bible, Gilgamesh becomes bestial, but he overcomes that reversion to nature and achieves the transcendental. In a way, only when Enkidu represents Gilgamesh's free and natural self is Gilgamesh a whole man. Without Enkidu, Gilgamesh has to act the beast himself, having nobody to play it for him. The legend suggests that no city man is wholly happy when he has lost the wilderness and the forest and contact with nature. The savage lies both outside the walls and within the yearnings of the ruler and the people of the city. Urban man dies with despair in his heart, unless he can range elsewhere.

The *Epic of Gilgamesh* is primal in the examination of the relationship between the savage and the city. It is the source of many of Europe's myths, it derives from the most ancient of urban civilisations, and it has an unrecognised power in folklore and psychiatry and even anthropology. The archetypes of Gilgamesh deal with the confrontation of the forest with the city, of the tree and the animal with humanity, of primitive man with urban man, of a hunting culture with a farming culture, of anarchy with authority, of instinct with self-control, and of the natural with the social.

In ancient China under the Hsia and Chou dynasties the same process of clearing the soil from the forest took place. The first large Chinese city near modern Anyang was centred in the dry loess soil of the Yellow River basin, more easily stripped of timber than the humid fertile earth of the Yangtze Kiang and Southern China. As one commentator has written, in China the progress of man has been marked by the retreat of the trees.[8] Through the assiduous cultivation of the centuries, the Chinese set up a civilisation of cities fed by irrigation canals and regulated by a calendar, that became more and more removed from the wild life of the nomads of the northern steppes and of the hunters of the western and southern mountains and forests. The Chinese alphabet, with principles which have hardly changed in three thousand years, was developed and inscribed on wood and silk; it served to perpetuate the power of a bureaucracy which every barbarian invader found indispensable. China early developed a society which could assimilate all conquerors. For it held itself to be 'the Central Kingdom' of the world with frontiers that defined the limits of civilisation. Therefore even occupation by barbarians was a digression. Chinese culture had its axis and turned all towards its way of life.

Frontiers are not so much lines as zones, through which the ideas of agriculture and civilisation may slowly reach the savage tribes of the beyond. China's history is unique in the building of a Great Wall on its borders as the expression of a philosophy.[9] In the Ch'in dynasty, the Wall was completed to mark the divide between the urban and the barbarian. The power of the Huns of the steppe with their horse mobility and love of plunder had threatened and would always threaten the Chinese emperors with invasion and the breakdown of society. While mountains and tropical forests and seas protected the other borders of China, no natural barrier lay to the north. Thus the building of a Wall by a great dynasty that gave its name to China also led to the finite construction of a mentality which still persists – the view of a world in which the limits between the civilised and the barbarian are exact and impassable.

Although ancient China was not exclusive and welcomed the influences of India and even of Greece and Rome, it was never to resolve the conflict between the mandarin within the city wall and the Mongol raging outside the gate. China's misfortune has always been a geographical one. However far its empire has expanded into the grasslands of Central Asia, it has never found a natural barrier against Hun or Turk or Russian, between the Chinese peasant regulated by the court and by the system of canals, and between the mounted savage of the plain, free to invade in his loose and shifting society. Both civilisation and climate have enforced the division of the Chinese from the peoples of the steppe. As the texts of Liao-shih and Ying-wei-chih declared: 'South of the Chinese wall is much rain and much heat. People make their livelihood by agriculture. They dress in silk or linen, they live in palaces or houses, they build cities and walls. In the great desert, there is much cold and much wind. People make their livelihood by herding and hunting. They dress in hides and furs; they migrate according to the seasons; cart and horse are their home. Thus, seasons and climate separate the south from the north.' In spite of a cult of a mythical beast, the dragon, representing natural forces, there was no end to the conflict between urban China and its periodic nomadic invaders. Even Gilgamesh could not have solved the problem.

The ancient civilisations of the Mediterranean, however, the cultures of Egypt and Greece and Rome, were bounded by the obstacles of the forests and of the deserts, with the central sea containing the widespread commerce of their influences. As in China, the civilisation of the Nile held those who were not Egyptians to be barbarians fit to be slaves; but its preoccupation was with the yearly cycle of birth and death, given by the sun and the silt of the great river. Although the conquering Pharaohs reached the Euphrates and the hot jungles of the Sudan, water and astrology were their obsessions, not the threat of forest or barbarian. The Egyptian attitude to its black southern subjects was one of patronage; the Sudanese were enslaved and represented in art almost naked with long angular limbs very like monkeys. Some of these Nubians received high office in the Egyptian court, but generally the blacker they were the more inferior they were thought. In the showpieces of Nubian tribute, the black man is represented by a row of decapitated heads on spikes or as the human foundation of the centre building. Like the Sumerians and the Chinese, the Egyptians considered themselves a people apart, the rest of mankind inferior.

Yet the Egyptians did not separate themselves from the rest of creation. It was a land of sacred cats and jackals and serpents, in which the gods and the Pharaohs were infused. Deities patronised special species and were embodied or represented in their shapes. There was 'the divine bull-dynasty of Apis, the sacred hawks fed and caged in the temple of Horus, Thoth and his ibis, Hathor the cow and Sebek the crocodile.' Many of these sacred creatures were worshipped in some parts of Egypt, yet killed and eaten in other places. It is the same in modern as it was in ancient India. Animals are hunted and eaten by unbelievers, yet venerated and pampered by the faithful. 'The sacred cow is not merely to be spared; she is a deity worshipped in annual ceremony, daily perambulated and bowed to by the pious Hindu, who offer her fresh grass and flowers; Hanuman the monkey-god has his temples and his idols, and in him Siva is incarnate, as Durga is in the jackal; the wise Ganesa wears the elephant's head; the divine king of birds, Garuda, is Vishnu's disciple.' And the bases of Hindu cosmology are the forms of fish and boar and tortoise in the Vishnu legends of creation.[10]

To the Pharaohs, even the ancient Greeks were barbarians, uncouth pirates and mercenaries from the north, and they were presented in Egyptian art in a comical way with hook noses and great bellies, bearing tribute. Yet the Greeks of classical ages, who had originally come from the mountains and grasslands of central Europe, only took to the sea because of the sparse offerings of their own mainland. While their population was limited in Homeric times, the hunger of their growing towns and the stripping of the wood from their own lands for their fires and fleets sent them out to found colonies and to bring home the timber and wheat which they needed to supplement the local produce of olives and vines. Like Gilgamesh, they had to voyage to distant and hostile forest lands, not for cedarwood for their stone temples, but for planking for their ships, since the Aegean was their life-blood. The Greeks always feared the sea, but depended upon it and fought the Trojans and the Phoenicians for control of granaries and mines as far away as Sicily. They accepted the institutions and prejudices of the ancient world, particularly in regard to slaves and barbarians; but they also produced philosophers who were sceptical enough to question an economic reliance on degraded human beings.

In Homeric times, when the warrior virtues were paramount, brutality in war was exalted and slavery was thought the just fate of all prisoners. It was certainly better than slaughtering them all as a sacrifice to the goddess of war – a contemporary Celtic practice. Yet by the time that Athens had created its complex urban society, based on trade, with a large body of free citizens and an occasional limited democracy, slavery became a matter of conscience for the first time in the ancient world. An Athenian tragedy, such as the defeat at Syracuse, when her surviving soldiers were worked to death in the Sicilian quarries just as Athens worked its own prisoners to death in the silver mines of Laureion, made even Plato consider the legal basis of man's inhumanity to man.

In an economy where manual labour was thought degrading to the citizen, who should be left free for the affairs of state, the abolition of slavery was unlikely. Plato's suggested slave laws were even harsher than those of Athens, which provided for the torture of any slave involved with the law courts, since slaves were held to be incapable of honesty. Although Plato could conceive that slaves had souls, he only condemned the enslavement of Greeks by other Greeks. He believed that the barbarians of the forest and the mountain should serve as Greek labour. His successor Aristotle also agreed that primitive peoples could only exist well as the servile instruments of free men. Aristotle hardly even wished to distinguish slaves from animals, complacently observing that 'with the poor, the ox takes the place of the slave'. In his *Politics*, however, he did concede that slaves had the power of reason, although they needed guidance by a master. Yet for a philosopher who himself might easily have been enslaved by the Macedonians, Aristotle could have considered more carefully the implications of classifying fellow men as no more than 'living tools' or 'animated property'.

Aristotle's justification of the slave by nature was to prove vicious not only to the Greeks, but also to the American Indians within the later Spanish empire. Yet immediately, Aristotle seems to have had little influence on his pupil, Alexander of Macedon. The great Hellenic empire created by that extraordinary conqueror left one enduring legacy – the concept that the division between East and West, between barbarian and civilised man, was unnecessary. Alexander was a founder of cities in the wilderness. He incorporated the peoples he conquered into a mixed Graeco-Persian way of life. He was the Gabriel of the inclusive Mediterranean civilisation, which Rome was to inherit. If he did not destroy the idea that barbarians lurked on the frontiers nor alter the divisions between slave and freeman, he did push back the limits of exclusion. The outlawed and the feared were now distant from the interior Alexandrian cities of the empire.

In fact, just as the word 'savage' was to become corrupted from meaning 'wooded' to meaning 'bestial', so the Greek word 'barbarian' lost its original meaning of 'non-Greek' and acquired its modern meaning. In Homeric times, when the pirate Greeks had been faced with the complex civilisations of Egypt and Persia, the term 'barbarian' had not implied any inferiority. It described someone who spoke a foreign language. Yet as the politics and the power of Greek imperialism grew, and as the Greeks after the sixth century before Christ began to call themselves 'Hellenes' rather than by the names of their original tribal groups, so their new national pride – hard-won in the wars against Persia – led to the demeaning of the word 'barbarian'. Its new significance was rough, alien, animal. In that sense, Euripides used it in the *Bacchae* as the cruel orgiastic counterpoint to the rational balance of the sceptical mind. Curiously, Euripides located the cult of Dionysos in the wilds of Macedon, just before Alexander was to make it the birthplace of the spread of Hellenic civilisation.[11] Yet in Alexander's distant Asian capital of Babylon, Athens itself was to seem a backwater – if not barbarian, at least less civilised than the imperial court in the East.

As the creators of many of our terms of philosophy and science and biology, the Greeks considered the relationship of men with the natural world as well as with other men. In Homeric times, the animals were venerated as attributes of the Gods. The father of the heavens, Zeus himself appeared as a bull to Danäe, while divine flying horses pulled the chariot of the sun. In urban Greece, however, there was the usual division between citizens and the living things of the wilderness. Dodona might have its sacred and prophetic oak, and Eleusis its mysteries in the groves. Eusebius might claim that an ancient Phoenician philosopher had declared: 'The first men consecrated the plants of the earth, and judged them gods, and worshipped the things upon which they themselves lived and their posterity.'[12] Hamadryads and wood-nymphs might still be thought to inhabit forests, while the Roman Ovid could relate in his *Metamorphoses* that Daphne grew into the laurel and was honoured by Apollo, or that Phaeton's sisters turned into bleeding trees which cried for mercy when their branches were cut. But the thickets were now being stripped for timber to make temples and ships and rafters, or else to burn as firewood. The shared vital force of blood and sap became a mere cult, while a feeling of being with green and growing things was altered to myth and folk tale.

The taming of the wild beast and the keeping of livestock by the Greeks near their cities led to a similar loss of belief in the divine attributes of animals. Domestication breeds superiority. The herdsman and the butcher thought of their flocks as meat or wool or leather, not as gods. And in a farming society, hunting was no longer a necessity, but a recreation. Even the goddess Artemis did not kill for food, but for love of the chase, as did the Greek heroes and the kings of Macedon,

Philip and Alexander the Great. More and more, the slaughter of wild animals became a pastime, while the keeping of tame ones was a matter of commerce.

Early Greek historians such as Herodotus combined observation and legend in describing the flora and fauna of other countries. Although the worship of trees and animals was passing in his homeland, Herodotus sought it abroad.[12] While a concern for other species was rare among Greek writers, Ossian defended the dolphin. 'The hunter of dolphins is immoral. That man can no more approach the gods with a welcome sacrifice nor can he touch their altars with clean hands. He who plans to destroy the dolphins pollutes those who share the same roof with him.'[13] And three philosophers, Pythagoras and Plato and Aristotle, were to defend animal creation more than they did their fellow men, thought to be debased. Pythagoras adopted the Egyptian belief in the transmigration of souls. When he saw a man kicking a puppy, he said the young dog incorporated the soul of a friend of his. He recognised his friend's voice in the yelp of pain. Plato also believed that animals had undying souls, although later classical philosophers only gave beasts a lesser spirit, an *anima* rather than the human *animus*. With his interest in biology as well as ethics, Aristotle inferred that man was a rational animal, and he saw the world as a process and a whole to be studied as an organic unity. The feeling of hunters for forests and beasts became the concepts of philosophers for those who could read and understand.

Plato and Aristotle did not question the degrees of men. The best thing was to be a Greek citizen, the worst to be a barbarian or a slave. But this did imply conformity and acceptance of the law: even Socrates had agreed to drink the hemlock after being judged in Athens. Indeed, Aristotle declared that the arrogant gifted man who could not fit into urban life should live outside the city 'like a god or a wild beast.' Superiority, like inferiority, should find its place in the woods, even if it meant joining the barbarians beyond the walls of civilisation.

With the breaking apart of the Alexandrian Empire and with the spread of the imperial and urban civilisation of Rome around the Mediterranean, the Greek use of the word 'barbarian' began to take on a wider significance. It came to mean what was foreign to the habits of the governing class, while savagery was what horrified the ruling morality. For instance, Strabo wrote that Posidonius was disgusted at the Gauls' habit of fastening the severed heads of their enemies to the necks of their horses; to him, it seemed both barbaric and savage, until he became used to the practice. On the other hand, the spectacle of brutality in the Roman circuses aroused little protest, being customary. At the

Games put on by Pompey in 55 B.C., however, the crowd cursed him, shouting against the killing of hundreds of elephants and lions and leopards. Cicero wrote that the people somehow felt pity 'that the elephant was allied with man.'[14] Plutarch, the Greek tutor of the Emperor Hadrian, took a sensibility for animals even further. He argued that men should extend their goodness and charity even to irrational creatures. They had a duty to domestic animals, to care for them in old age, not treat them like old shoes or dishes and throw them away when they were worn out. And wild animals should only be killed for meat which was necessary, not for a delicacy. Flesh should only be eaten for hunger, and beasts never killed in Games or for the sport of the chase.

Plutarch was not heeded. The Emperor Trajan made him a consul, then had eleven thousand wild animals slaughtered in a single day to celebrate a triumph against the Dacians. Later the Emperor Commodus himself dispatched panthers and rhinoceros and running ostriches with his unerring arrows, but from a safe place. As Gibbon wrote in his *Decline and Fall of the Roman Empire*, 'In all these exhibitions the surest precautions were used to protect the person of the Roman Hercules from the desperate spring of any savage who might possibly disregard the dignity of the Emperor and the sanctity of the god.'[15]

The Romans had corrupted the Greek theatre into a sport of multiple murder. In some plays, slave actors were even killed on stage for the sake of realism. For the Greeks, indeed, who had thought themselves wholly civilised, their defeat by the Romans led to a feeling of inferiority. If originally they had come to Rome as teachers on the principle 'When in Rome, do as the Greeks do', they ended as servile imitators and slaves. While the early Roman republic had first met Hellenic civilisation as a simplistic military power, the later Roman Empire was sophisticated beyond the imagination of an Alexander. Juvenal's portrait of greedy degenerate Greeks, magicians or quacks with quick wits and unlimited nerve and absent morals, was as dismissive as any attack on despised aliens. The very luxury of Rome led to the creation of the myth of the noble savage by Tacitus, who reversed the standards of civilisation so that forest life seemed to be good and the city evil.

In his *Germania*, Tacitus praised the barbarian peoples who had decimated the Roman legions probing beyond the Rhine. To him, the Germans represented the simple virtues of courage and chastity and frugal living that had once led Rome to break through the tangled Appennines and conquer Italy on its way to imperial glory. Tacitus implied that urban life destroyed morality. He approved of the spread villages of the Germans, built of undressed timber wherever a spring or a plain or a grove had taken their fancy. If their children grew up naked and dirty, they also grew up strong and lusty. Even German slaves had their own houses and plots of land; they acted more as tenants than bondsmen, while the master's wife and children looked after his house without benefit of slaves. Usury was unknown, pride in the family and tribe and battle was all. So Tacitus invented the myth of the noble savage to serve as the mirror of urban iniquity.

Such a picture of the primitive Germans was, naturally, a gratifying contrast to the squalor of Rome. Yet when these barbarians actually left their woods and broke through the borders of empire and sacked Rome and recolonised Italy, they proved more greedy and vicious than the Romans themselves. The virtue of the barbarian lay in his distance from civilised man; the nearer he approached, the more savage he became. St Augustine, mourning the approach of the Vandals from his diocese in North Africa, saw no good in their gift for destruction. However corrupt Rome might have been, it did preserve the minimum conditions necessary for civilisation and worship.

The rise of two world religions based on the ancient cultures of the Mediterranean, Christianity and Islam, added more definitions to the distinctions between the savage and the civilised. Christianity sprang from Judaism, which had a long tradition of choosing the liberty of pastoral life rather than the tyranny of the city. When Moses led his people from slavery in Egypt, he led them into forty years in the wilderness of Sinai before finding a land of milk and honey. There the barbaric Jews attacked the little cities of their promised land, Jericho and Gaza and other alien walls, and when they were enslaved and deported again, they wept by the waters of Babylon. They set up their own holy city at Jerusalem; yet when it was conquered by the Romans, the zealots withdrew to the barren places around the Dead Sea. In the wilderness and in dispersion lay the final freedom of the Jews and the visions of the Prophets, who only came to the wicked cities of the plain to threaten them with fire and brimstone and desolation.

Jesus Christ was born in the flight from the murderous city and He was brought up in pastoral surroundings. He withdrew to the wilderness to achieve spiritual strength, and He only returned to Jerusalem to be sacrificed on a crossed tree. The manger was His birthplace, shepherds His barbaric nurses. The city was His civilised place of judgement, the barren rock outside its walls His tomb and resurrection. Although He preached the virtues of the meek, He knew the need of the wild for the renewal of the spirit. He was truly a prophet in the late tradition of the Jews.

It was the flaw in the Church of Christ that it became too identified with the metropolis and its power. St Paul made Rome the first centre of the new established religion, Constantine made the new Rome of Byzantium its second centre. The careful counterbalances which the Gospels preached of the need for city and country, of entry to Jerusalem and withdrawal into wilderness, were lost in the hierarchy of urban control. Those later prophets like Jerome who withdrew to the desert to mortify their bodies and purify their faith had little direct impact on the Church. Their reversion to a savage and simple way of life seemed praiseworthy, but excessive. To live like a beast in a cave was surely less good for scholarship than to study in Alexandria or Byzantium. Saints in the sands might be the seeds of faith, but the Church in the metropolis was its brains and sinews. The spread of Christianity under the Byzantine Empire encouraged the divisions of humanity. The late Roman Empire had tried to include the invading barbarian tribes by offering them citizenship. Yet, if Christianity brought hope to the large slave population of the Mediterranean by preaching equality in the sight of God, it also yielded up pagans to a damnation worse than barbarians, to an everlasting torment. By defining moral practice strictly, Christianity became the arbiter of sinful humanity. All those who were not Christians were condemned, whether civilised or barbaric or savage, whether urban or agricultural or nomadic.

In his *City of God*, the perfect vision in heaven of an imperfect Rome on earth, St Augustine did much to equate Christianity with urban life and paganism with existence in the wilderness. Even if such distinctions became less significant after many barbarian tribes were converted to Christianity, yet the confusion of the Church of God with the city remained, for to be a Christian after Constantine was to be a Roman.[16] Although the leading power in Christianity was to become Byzantium rather than Rome itself, yet it was the New Rome, the champion of a militant and exclusive religion set against the older rituals of the Europe of the forests and the mountains.

Christianity never quite destroyed the ancient link between men and trees. This primitive consciousness was later explored in depth by the seminal anthropologist Sir James Frazer. He inferred that Virgil's 'golden bough', plucked by Aeneas from the sacred oak at Nemi, was mistletoe, also holy to the Druids. He combed the world for examples of rituals commemorating the cycle of the seasons, the yearly birth and death of green things. To him, the priest of the grove at Nemi, who was King of the Wood only as long as he could defend himself against all attackers, was symbolic of the harsh renewals of nature. Frazer understood that the numerous ceremonies designed to ensure the fertility of the earth had begun in belief, if they were to end in custom. Where the Adonis myth had once led to the human sacrifice of a youth each year in classical times, the maypole dances of European midsummer were to become no more than festivals, and the execution of the Green King or Wild Man in German villages was to be finally symbolic, a mere lopping of a wicker head plaited with leaves. As men became less savage and more civilised, so they would substitute a show for a sacrifice – bread and wine for the body and blood of the ritual victim of the cycle of life and death.

Frazer also pointed out that 'to the savage the world in general is animate, and trees and plants are no exception to the rule.' The savage presumed that growing things had souls like his own, and he treated them accordingly. Even the classical vegetarian Porphyry had observed that primitive men led an unhappy life, since their superstition did not stop at animals, but extended even to plants. 'Why should the slaughter of an ox or a sheep be a greater wrong than the felling of a fir or an oak, seeing that a soul is implanted in these trees also?'[17] By the law of the tree-worshipping ancient Germans, indeed, the man who peeled the bark off a living tree had his navel cut out and nailed to the wound in the trunk, after which he was forced round and round the tree until his guts bandaged his crime.

Tree worship was not unreasonable when Europe was mainly forest. When Julius Caesar had reached the Rhine, he had found Germans who had travelled east for two months without reaching the end of the woods. Early villages in northern Europe were little clearings in an immensity of trees, while even the Po Valley was thick with elm and chestnut and oak. The Celts actually worshipped tall oaks as the images of the Creator, and their ceremonies and governments were held in oak groves. Forest peoples took wood to be holy, the maker of shelter and of fire through friction, and when their trees were no longer the places of their rituals they built their new temples traditionally. So the cathedrals of later Gothic Europe were given aisles of stone pillars that rose to interlace in rib vaulting overhead like the branches of a petrified grove.

Unlike Gilgamesh, the Christian myth of creation gave mankind dominion over it as well as the duty to praise God's handiwork. It is interesting that *Genesis* stated God made grass and herbs and seeds and fruit-trees on the Third Day of Creation, before calling up the sun and the moon and the heavens. Fish and fur and fowl had to wait until the Fifth Day to be given flesh, while man was made in God's image (as well as a female) on the Sixth Day before the divine day of rest at the end of the week. God's blessing and orders to the primal Adam were clear: 'Be fruitful, and multiply, and replenish the earth, *and subdue it: and have dominion* over the fish of the sea, and over the fowl of the air, and over every living thing ...' Adam was given every herb bearing seed and every tree with fruit: 'To you it shall be for meat.' God also granted to all beasts and fowls and creeping things 'every green herb for meat.'

In *Genesis*, God the Creator told the first man to breed and fill the earth and to subdue it, and then to change it to his own purposes. He was given dominion over every living thing, wild or tame. He was also given for meat all herbs and trees that were useful to him. Animals were only granted green herbs for meat. When the first man and the first woman, Adam and Eve, fell from the natural paradise of the Garden of Eden because the subtle serpent gave Eve the apple from the tree of knowledge of good and evil, God's commands to human beings altered. Eve would now suffer in childbirth and be ruled by her husband. Adam would toil in the field as a farmer and would labour from childhood until death. The only gifts that God gave the first man and woman before driving them forever from His Garden were coats of animal skins rather than their chosen aprons of leaves. They would have to live as hunters and herders to clothe themselves as well as tillers of the soil to get their bread. Their first two sons, Cain and Abel, were a ploughman and a shepherd.

The strictures in this parable of the creation of the Christian world were reinforced by a new version of the old myth of the Flood, which dated from Sumerian times. Although mankind had multiplied and turned to wickedness, God decided to destroy His animal creation along with erring humanity. He told the one just man Noah, 'The end of all flesh is come before me; for the earth is filled with violence through them; and, behold, I will destroy them with the earth.' But Noah was ordered to save at least two of each sort of flesh by building a large ark, stocked with all varieties of fodder. Thus God made a man the conservationist of all the species. When the flood waters rose, Noah and his family lived with and fed the whole animal world, which depended on them for survival. The sign to Noah that the waters were drying up was an olive leaf in the beak of a dove, a bird bearing plant life. But once Noah had beached the ark on the slopes of Mount Ararat and had released all the creatures, God assigned to him power over those he had saved, saying, 'And the fear of you and the dread of you shall be upon every beast of the earth, and upon every fowl of the air, upon all that moveth upon the earth, and upon all the fishes of the sea; into your hand are they delivered. Every moving thing that liveth shall be meat for you.' But the savages which killed men, whether human or bestial, those must be punished. 'Whoso sheddeth man's blood, by man shall his blood be shed: for in the image of God made He man.'[18]

These divine commandments made human beings the saviours and exploiters of all animals. They had a right to prey on wild beasts, which might prey on them. And domestic animals were for their riches and enjoyment. Before Job was struck down by his tribulations, he was wealthy in livestock – seven thousand sheep, three thousand camels, five hundred yoke of oxen and the same number of asses. These were far in excess of the needs of his family. The Old Testament did recommend the care of farm animals, which should be rested with mankind on the Sabbath day. As *Proverbs* declared, 'A righteous man regardeth the life of his beast.'

Yet the Hebrew faith was set against the pagan worship of animals and plants, particularly the Near Eastern veneration of bulls. Moses had to cast down the Golden Calf and its rites before he could lead the Jews to the Promised Land. In two episodes in the Gospels, Jesus Christ appeared to value human beings far above domestic creatures and certainly trees. As St Augustine commented in his attack on pagan faiths, 'Christ Himself shows that to refrain from the killing of animals and the destroying of plants is the height of superstition, for, judging that there are no common rights between us and the beasts and trees, He sent the devils into a herd of swine and with a curse withered the tree on which He found no fruit.' One man's sanity was restored by Christ at the expense of a whole herd of Gadarene pigs, a tree was blighted because it was bare. Augustine went on to explain that the moral rules that the Saviour laid down for the behaviour of men to each other did not apply to their treatment of animals and green things.[19] Jesus had confirmed God's grant of dominion over nature to Adam and to Noah.

There were traditional saints who did show respect for other varieties of life, which often sustained them in return. Ravens were said to have taken bread to Paul the Hermit, swans looked after Cuthbert, a lion guarded Gerasimus after he had removed a thorn from its paw. In medieval times St Francis of Assisi was to preach a love of animal creation, based on the assertion of Jesus that not one sparrow was forgotten before God. And even the chase acquired its patron, St Hubert, who stalked a stag on Good Friday, only to see the cross shining between its antlers and to hear the voice of Christ saying, 'Unless you turn to the Lord, Hubert, you shall fall into Hell.' This vision did not prohibit hunting except on holy days, but it did place the crucifix among the horns of a deer. The whole of creation was divine, after all, even if given into the hands of man.

It was more usual for saints to agree with Augustine and demean plant and animal life for fear that these might be worshipped. St Boniface cut down a sacred Hessian oak and used its wood for the rafters of a chapel for St Peter. In early religious paintings, the Devil and his imps were represented in the shapes of animals, flying serpents and swine, wolves and bears. Augustine taught Christians to deny wild things and to look towards the City of God. But another rising religion in the Mediterranean, Islam, took the opposite view. Its success lay in its purity and its plain and simple rules, which appealed to those who lived close to survival between the stars and the sand. Without falling into the causalities of an Aristotle or a Montesquieu and claiming that countries derive their governments from their climates, it is arguable that the conflicts between the people of the forest and of the plain, the people of the interior and of the coast, the people of the city and of the desert have much to do with environment as well as economics.[20] Six centuries may divide the Arab historian Ibn Khaldun from the French historian Braudel, but both would agree on the reason for the success of the Muslim faith. Islam was 'the desert, the emptiness, the ascetic rigour, the inherent mysticism, the devotion to the implacable sun, unifying principle on which myths are founded, and the thousand consequences of this human vacuum. In the same way, Mediterranean civilisation grew up under the determining influence of the emptiness of the sea; one zone peopled by ships and boats, the other by caravans and nomad tribes. Islam, like the sea and like the desert, implies movement.'[21]

A place influences its people, new or old. The invader becomes the settler and finally the native. Cultures derive from their countries. At bottom, as Braudel says, a civilisation is attached to a distinct geographical area and this is itself one of the indispensable elements of its composition. Another element of a civilisation is its exclusion from power and influence of those peoples who do not share its language, beliefs and boundaries. The prophet Muhammad had founded Islam by a unique act of exclusion and aggression, more sweeping than any Chinese contempt of the barbarian or Christian condemnation of the pagan. When he formed his first religious community or *umma* at Medina after his withdrawal from Mecca, he divided humanity into the *Dar al-Islam* and the *Dar al-Harb*. Those who did not dwell in the Abode of Islam – and these only numbered a few hundred at the time – dwelt in the Abode of War. It was not a matter of stating that those who were not for Islam were against Islam; it was a declaration of holy war against all unbelievers. Muhammad's attacking faith inspired a divided group of tribes, the Arabs, into acquiring a Mediterranean empire in less than a century. As Ibn Khaldun noted, a Christian then could not even float a plank in that sea. Only in Paradise after death was a garden of trees and plants, created by Allah, to be enjoyed by all true believers, while the Koran said little about the treatment of animals except to prohibit the eating of pork.

During the 'dark ages' of Europe, when the invading forest tribes lost the urban habits of the Roman Empire and when the farms of the colonists reverted back to the woodlands, Arab civilisation reached its height. Ibn Khaldun may have imitated Tacitus in extolling the warlike and savage virtues of the wandering Bedouin and Berbers in contrast to the urban vices and luxuries of the ruling Arab families, who were slowly weakened until they 'fell into a long sleep in the shadow of glory and peace.' Yet the fact was that the extraordinary achievements of the Arabs in medicine, chemistry, physics, botany, astronomy, mathematics and, above all, in clinical observation could only have been fostered by these enlightened urban centres and courts. The Christian counter-attack on Islam which was called the Crusades was a barbaric assault on a cultured people, who both beat off and instructed their attackers.[22] Without Arab science as the intermediary between Greek method and Renaissance enquiry, the technological explosion of Europe would have been impossible.

In spite of the religious fervour and rapacity which inspired the Crusaders, divine greed could not sustain them in the Holy Land. There were too few of them on the ground, while many of the few could not tolerate eastern conditions and parasites. They existed, anyway, on the disunity and the tolerance of the Muslims. When Saladin brought his peoples together and recaptured Jerusalem, the sacred purpose of the Crusades was over. Some Franks in armour and on horses had held four principalities in the Near East for a century or so. They were out of place, unable to colonise. The Muslims were in their homelands. 'The weather was natural to them,' Richard of Devizes noted, 'the place was their native country. Labour was their health, frugality was their medicine.'[23] Malaria and cholera and typhoid decimated the western invaders, as well as the illnesses caught from crowding together in castles and walled cities – smallpox and measles, whooping-cough and venereal diseases. Microbes killed the Crusaders in greater numbers than the Muslims did.

The Frankish cavalry taught their enemies the value of the horse.

Encased in mail and mounted on his charger, one knight could overwhelm a hundred foot-soldiers. A cult of the horse arose. It is echoed in God's words to Job, translated at the beginning of the British colonial epoch in America.

> The glory of his snorting is terrible.
> He paweth in the valley, and rejoiceth in his strength:
> He goeth out to meet the armed men.
> He mocketh at fear, and is not dismayed;
> Neither turneth he back from the sword.
> The quiver rattleth against him,
> The flashing spear and the javelin.
> He swalloweth the ground with fierceness and rage ...

Only a few hundred Spaniards on their horses were to overthrow the Indian empires of Mexico and Peru. Only the Indian adoption of the horse was to make the Sioux and the Apaches the best light cavalry in the world. If the Crusaders lost on land in the Near East, they left a memory of the noble beast which they rode, and they took with them the Arab tools for founding other colonies over the oceans, particularly the sternpost rudder and the compass. Also through the Norman kingdom of Sicily, Greek science and medicine developed by Muslim scholars entered the intellectual awareness of backward Europe. For the Crusaders themselves were illiterate. Reading was not necessary for a human war machine. Priests could do that in Latin.

The assault of Europe on Islam was defeated, but hardly. Ibn Khaldun was proven correct. The Arabs were 'all softened, even the cutting edge of their swords.'[24] Civilisation is too often the pander of its own downfall. If the early Islamic conquerors were fierce and ruthless in their nomadic assaults across the deserts of the Near East and North Africa, the later Arab rulers were to be destroyed by the savage hordes of the Turks and the Mongols. It is interesting that, at one point, the Arabs actually defeated the T'ang Empire at the Talas River and forever prevented the expansion of China, while converting Central Asia to Islam. Yet they had no forests as the North Europeans had to defend them from the Mongol horsemen, when these came across the steppe and over the mountains to reduce many of the cities of Asia to dust and total massacre. Genghis Khan, indeed, so hated urban culture that he had to be dissuaded from returning the whole of northern China to pasture for his war horses. Even Ibn Khaldun would not have praised such warrior virtue.

Yet Arabic civilisation was to endure, even in its decline, because of its pure faith and policy of toleration. Its missionaries spread the Koran through Africa and Asia, while its local rulers accommodated religious dissent and strange practices. The evil of Islam lay in its Mesopotamian inheritance – a heavy reliance on slavery and the expansion of the slave trade in Africa for commercial profit. Although slaves could attain positions of great power in the courts and armies of Islam, the principle of slavery was retrograde even in barbaric Europe, struggling towards a legal definition of individual rights through early feudalism and serfdom. If scientific and medical research was the great gift of the civilised Arab to the medieval European, the practice of enslaving the despised African for profit was his curse. The long Arab trade routes and outposts that reached as far as China and Indonesia and East Africa were the forerunners of the sea-empires of the Iberian powers; but the slaving caravans to the Sudan and West Africa were equally the scouts of the most violent misuse of mankind by men in European history.[25]

In feudal Europe, there were many serfs, but few slaves. The Christian religion had flourished originally among the slaves of the Roman Empire. Its doctrine was attractive because it preached the equality of souls. In the sight of God, all human beings were of equal worth. Their future in heaven depended on their good works on earth. But when St Thomas Aquinas discovered Aristotle and introduced his philosophy into the Catholic faith, he followed the Greek philosopher's feeling about the rest of the animals which were not rational. 'Beings that may be treated simply as a means to the perfection of persons can have no rights and to this category the brute creation belongs,' Aquinas wrote. 'In the Divine plan of the universe the lower creatures are subordinated to the welfare of man.' Aquinas did continue to state that mankind should 'respect and obey the order established by the Creator' which included fair play to beasts. But these only existed for the well-being of Christians.

The nature worship of the Far East was also being opposed and superseded by Buddhism. The first Buddhists disputed whether trees had souls, but they finally decided that green things had no minds or feelings. Spirits could inhabit them, and souls might transmigrate into them. Buddha himself told a parable about a genius residing in a tree, which cried out to a carpenter about to cut it down, 'I have a word to say, hear it!' The Bo tree under which Buddha lived is still sacred, but as a shrine, not as a totem. As for animals, Buddhism respected their lives because souls might have migrated into them; but beasts were not held to possess their own souls, and they did not deserve human prayers.

Buddhism incorporated animism and nature worship into its veneration of the whole of creation as a way to praise the divine.

This summary, which aims to introduce certain themes in the rise of civilisation in the Middle East and Asia and Europe, has concentrated on the historical change from a savage or barbarian society to an urban one, which then exploited primitive peoples as slaves to build up the reserves of foods and weapons necessary for city life. Such a process was first a matter of clearing the trees from the soil, so that it could be used for crops. The natural prejudice of each primitive tribe that it was *better* than other tribes was elevated by the urban civilisations into an ideology of innate difference and superiority, which could excuse the institution of slavery. The savage and the animal and the seed were defined as only fit to serve. This assertion was strengthened by the rise of Christianity and Islam, which promised salvation only to those who worshipped their sacred texts, the Bible and the Koran. They taught the ultimate weapon for the definition and control of the savage, both within and without man – the holy word.

For it is language that finally distinguishes a man from a beast, and one civilisation from another. Both Christianity and Islam particularly used words to defeat the pagan cultures of the wilderness. The texts of the Bible and the Koran were written and certain, while the cults of the Druids and the Africans of the forest were oral and forgotten with the destruction of their societies.[26] Written language is the chief separation between urban men and the tribes which they seek to convert and dispossess. Words are the devices by which priests and administrators net men from the wilderness for their service. As Prospero boasted to his slave Caliban:

> When thou didst not, savage,
> Know thine own meaning, but wouldst gabble like
> A thing most brutish, I endow'd thy purposes
> With words that make them known.[27]

Caliban admitted the gift of language, yet he said his only profit from it was to learn how to curse his new master who had taken over his island. The words that fitted his mouth so badly were merely sounds taught to him to prove that he was rightly enslaved.

Paul Radin has pointed out that primitive man may be a philosopher without a pen. Two things only are common to all aboriginal peoples, their lack of a written language and their possession of the land which other men want. The fact of written thought, the magic of the alphabet, is one of the chief reasons for their dispossession. In the beginning of Mediterranean civilisation was the *written* word, and it could be worshipped. In Judaic mythology, the Hebrew letters disputed with God Himself as to who should have the honour of originating the Bible. In a mutable and everyday world, the stability of the written word, its power to record and to prove the past, became a weapon in the minds of churches and civilised men. They believed that the word itself might be real, that what was visualised and inscribed was true. Thus the biblical fable of the progenitors of the human race, the three sons of Noah, would be used to prove the righteous superiority of Japhet's Europe and Shem's Asia over Africa, where Ham's descendants were condemned like Caliban to be the inferior servants of the rest of mankind. For 'God shall enlarge Japhet and he shall dwell in the tents of Shem,' while Ham 'a servant of servants shall be unto his brethren.'

Yet for primitive man, the two great realities other than himself were the realities of the world of the senses and of his society. For him, thought was only one mechanism for adjusting and controlling conflicts between the individual and his fellow men and nature itself. Intuition and feeling and dream and totem were equally important in reconciling a person with his place in life. The Ojibway Indians of North America give their marrying clans the names of animals, the wolf and the bear, the deer and the rabbit, and the members of the clan are thought to share in the natures of their counterparts, who may even help them to the afterlife. As a Tlinglit poem from Alaska runs,

> If one had control of death,
> It would be very easy to die with a wolf woman,
> It would be very pleasant.

As for words, they were merely spoken language to the primitive mind, a description of the world unless they were an incantation or an oral tradition. Yet as Caliban discovered, the spoken and written words of Christian Europe were a trap, binding on the past and future although nature itself was seasonal and changeable. The Christians gave the subjective word and thought an objective and rigid value, which warped their view of themselves and the world. In Radin's words, 'the distortion in our whole psychic life and in our whole apperception of the external realities produced by the invention of the alphabet, the whole tendency of which has been to elevate thought and thinking to the rank of the exclusive proof of all verities, never occurred among primitive people.'[28]

Thus the savage man could not understand the concept of 'the savage'. To him, wood and stones and sky and wilderness were animate, while he shared in the being of beasts. He was at one with them. Being born of nature, he was his nature. No written words stood between him and his perception of the earth, only the memories of the tribe that begat him. When he met Christians and Muslims with their objective concepts and fundamental beliefs, he found them unnatural and called them gods at first, and devils later. Although he might learn to communicate with them in his language or theirs, he could not generally explain his intuitions of the world.

Chomsky may assert that all languages have common principles underlying their structure and that speaking is biologically determined, and Lévi-Strauss may state categorically: 'He who says man, says language, and he who says language, says society.'[29] Yet the mind of some intelligent men is not always capable of expressing itself in clear language, which does not mean that they may not contribute to human society. Although Lévi-Strauss admits that the first speech was in poetry and that reasoning was invented long afterwards, yet the conflict between primitive languages and those capable of logical arguments has generally led to a total misunderstanding which has often resulted in massacre and eviction. Not so much a war of words but a confusion of the unintelligible has caused the destruction of the savage and the barbaric in their encounter with the civilised.

Words have long been the weapons of dispossession and alienation. Misunderstood, they have stripped a primitive man of his life and land. Once learnt, they have severed him from his bond with the natural world. By setting down a logic of thinking, they have amputated the sympathies of feeling. The animism which once linked aborigine and nature has been called a confusion. Categories have fenced all out – this God alone – that a human, that a beast – those the trees, which are plants and not spirits. When all the earth was savage, no man knew that he was so. He had first to be told, to hear it.

Once the savage, like Adam, had recognized his nakedness, he had to clothe it. He had to leave the hunt for the farm or the shanty in the city. The forest that had encompassed men and had awed them into worship had to be transformed by them into the material for their lives. As they moved from the gloomy existence of hunters in the backwoods to the more open life of farmers and herdsmen, they cleared away their old environment and recycled it for fire or shelter or tools. The psychology of forest man gave way to the psychology of field man, as the timber retreated before the axe. In that new way of life, in that removal from the dictates of the wild trees to the safer confines of the cabin and the house, men also put a distance between the brutal, but restrained behaviour necessary for survival by hunting and the milder behaviour possible for agricultural folk who could store food and plan a seasonal diet. As Gilgamesh knew, culture became more possible at each remove from the wilderness and more critical of that primitive past.

The unknown is usually the enemy, while the misunderstood is always the savage. As the medieval world slowly pushed back the wilderness to make fields, as it became more educated and urbane, so complacency entered the citizen, as it had entered the Mesopotamian and the Egyptian. In the most advanced of societies before the European Renaissance, the Great Wall stood as the actual divide between culture and savagery, and Shao Yung could declare: 'I am happy because I am a human and not an animal; a male, and not a female; a Chinese and not a barbarian; and because I live in Loyang, the most wonderful city in the world.'[30] So, at the first beginnings of global expansion by sea, traditional attitudes were ready to condemn the alien as bestial, especially alien animals and women and primitive tribes and peasant economies. To be human, one had to be an urban man, far from the wild woods and the beings within them.

2
THE AMERICAN SAVAGE

'THE FLEET WAS a thing most wondrous to behold. In the morning it was like to a forest which had lost all its leaves and fruits, and lo! of a sudden it was changed into a magnificent orchard, resplendent with green leaves and flowers of varied colours; for the standards and the flags were innumerable ...'[1]

So the chronicler of King Dom João the First described the ships massed for the conquest of Ceuta in 1415. This Arab city was to be Portugal's first stronghold in Africa, a base for its thrust across the globe to win Europe's original seaborne empire. The caravels with their three masts and triangular lateen sails were made from the timber which still grew on the mountains of the Algarve, so that the floating forest seen by the chronicler was translated from the earth to blossom with standards heralding the crusade of Catholic traders and colonists against the Moor and the heathen. The trees of Europe were cut into service to carry Christianity to the savages of Africa and the Far East. For, as the first reports of Inner Africa from Antonio Malfante stated in 1447, the black tribes there were 'in carnal acts like the beasts' and 'eaters of human flesh.'

Although the medieval Portuguese seemed only semi-civilised to refined Arabs and wholly barbarian to the Chinese, who condescended to these 'sea-robbers', their voyages and discoveries were to set the aggressive urban civilisation of Europe against the forest cultures of West Africa and later of the Americas. In this shock, European imperialism was to be borne on the black gold of slavery and the yellow gold of commerce, defended by the walls of wooden ships and colonial fortresses. These outposts of Europe, moving across the ocean or fixed in the fringes of continents, encouraged and paid for the expansion of the small cities of the motherland. In the way that Venice had grown through the trade of the Levant from some marsh villages into a stone trading metropolis set on logs in its lagoon, so Sagres was to rise on the commerce of Africa and the Spice Islands, while Cadiz was to flourish on the profits of the Americas. When the thrust of the Portuguese faltered in the sixteenth century as the power of its greater neighbour of Spain overtook it, the wealth of the sea coast rebuilt the cities of the interior of the Iberian peninsula, which could never have developed on their own barren resources.

The capital and financial centre of the new kingdom of Castile was Valladolid. The first business of the growing city was to make a little dominion round its walls to feed itself. The gold and silver from the New World paid for its rooted and blossoming orchards to be irrigated by the Pisuerga river, while vineyards and olive groves could now be financed on the dry pastures of the hillsides, which could hardly support the grazing of their lean sheep and cows. Pinewoods were planted to replace the old trees, long since stripped by goats and rendered into charcoal or into beams for the noble mansions which were the envy of such travellers as Pedro de Medina. In Valladolid, slaves from Africa and Mexico waited at the tables of the grandees, and the Catholic faith burned purely there. The Plaza Mayor saw two great auto-da-fés of heretics flaming for the beginning of the Counter-Reformation.[2]

This powerful little city also saw another crisis of conscience in the year 1550. The king had ordered some of his councillors to listen to a great debate between the leading defender of the American Indians and the apologist for their oppressors. This early self-criticism within a small country, which had founded Europe's second global empire, was to affect the minds and behaviour of governments and colonists during the imperial centuries that were to come. The questions were insoluble. By what right did one people take the land of another? Did civilisation justify oppression? Was any war so just that it excused massacre? Were primitive men natural slaves? Did savages have souls to be saved? Or were they at the level of useful animals? These questions had been asked since the men of the farms had first resisted the men of the forests, or the villagers had defended themselves against the highlanders. Now the fact of the conquest of a vast empire in the Americas, inhabited by Aztec city-dwellers as well as cannibal Caribs, provoked a debate of sinuosities and contradictions as labyrinthine as the marshes of the new-found Mississippi delta.

The Spaniards of that time were still fresh from the harsh local life of medieval days. Huizinga opened his masterpiece, *The Waning of the Middle Ages*, with the observation that all things seemed more clearly marked at that time. The contrasts were greater, the calamities and the honours, the sufferings and the faiths, the miseries and the riches. The early Spanish conquistadors were still crusaders in their approach to the New World. Their passions and contradictions were extreme. They lived in 'that perpetual oscillation between despair and distracted joy, between cruelty and pious tenderness' which characterised medieval life. Their belief in the ritual of possession and in the justice of execution and torture was also medieval. Their plunges from victory to rout, from fortune to ruin were expected, as were their excesses of revenge and mercy. 'So violent and motley was life, that it bore the mixed smell of blood and of roses. The men of that time always oscillate between the fear of hell and the most naïve joy, between cruelty and tenderness, between harsh asceticism and insane attachments to the delights of this world, between hatred and goodness, always running to extremes.'[3] According to Huizinga, after the close of the Middle Ages, the mortal sins of pride, anger and covetousness never again showed the unabashed insolence of the preceding centuries. He did not take into account the conquistadors.

The Portuguese and Spaniards approached the unknown continents of Africa and the Americas with greed and prejudice as barbaric as that of any native tribe. They had a lust for the yellow of gold which was direct and primitive. Equally basic was their bigotry about strange peoples. In legend and art, the unknown parts of the world were inhabited by demons and beasts and monstrous men. These fantasies became gargoyles on the gutters and under the choir-stalls of the Christian cathedrals. Even after Columbus had landed in the West Indies, John of Holywood could describe the natives of the New World as 'blue in colour and with square heads.'[4] To the invading Spaniards, the Indians of the forest were similar to the wild men from the woods which the mummers played in the streets of the medieval towns. These were dressed all in leaves and carried giant clubs. In the morality plays, they represented the rude and brutish life of the forests that had been cleared to make the fields and villages of Europe.

The problem was that the denizens of the woods, human or brute, were thought only fit for the chase. If not killed, they became captives and served. As John of Salisbury had noted in his *Policratus*, the nobles and the knights pursued wild beasts with greater frenzy than they did the enemies of their country. 'By constantly following this way of life, they lose much of their humanity and become as savage, nearly, as the very brutes they hunt.' Such rare criticism did not affect the ruthless pursuit of most wild animals from rabbits to boars and bears by all members of medieval society. Those few which were sometimes a threat to human beings such as packs of wolves were the most cruelly treated.

In France, a trapped wolf might be flayed alive and then mutilated, as if it were a demon to be exorcised. Unfortunately, religious images presented devils as animals, and the current medical practice of casting out demons to cure hysteria and nervous diseases led to actual identification of beasts of prey as agents of Lucifer. To Norsemen, the raven had been the bird of death, but to the medieval villager, the fox and the hog were actually imps and evil forces. Satan wore the face of a beast.

Although human play has been recognised as part of culture, it was unfortunately applied to the pursuit of wild things, as it still is.[5] Hunting was designated as a sport, the torture of captive animals was used as a circus or a drama – the pit of our theatres derives from the bearpit. The very quarry was called 'game' or 'wild game'. A salmon was 'played' as a cat 'plays' with a mouse. In the era of carnage of the Middle Ages, necessary hunting to provide meat became organised ritual slaughter to provide entertainment for the privileged. As late as Tudor times, Queen Elizabeth the First would herself dispatch driven deer with a crossbow and cut the throat of a hart or take off its ears with her own hand. Aristocratic chases of fur and fowl and fish began to be seen as recreations, even as amusements. The victims were less for the larder than for the number and the exercise. Man at play was a worse killer than man in need of food.

At its worst level, the baiting of wild beasts, particularly bulls and bears, degraded the spectators beneath the objects of their cruelty. Mastiffs were loosed on their chained victims. The crowd rejoiced in blood-lust and grew affectionate over the survival of great fighting creatures such as Shakespeare's Sackerson – although the bard could also show his terror of the beast in *A Winter's Tale* with its famous stage direction in Bohemia, *Exit, pursued by a bear*. The Queen herself preferred the baiting of beasts to play-going. It was a more virile and traditional entertainment. Dogs and cats, too, were often tortured by the people, particularly the black cat, thought to be the familiar of witches. And birds were blinded, the linnet and the thrush, so that they might sing more often in their cages.

This was the attitude to the rest of creation which the European seafarers brought to the savage worlds of Africa and the Americas. As always, an occasional voice spoke up against the prevailing attitude. In his New World, Sir Thomas More made the Utopians relegate the pursuit of game to their bondmen. 'For they count hunting the lowest, the vilest and most abject part of butchery ... the hunter seeketh nothing but pleasure of the silly and woeful beast's slaughter and murder.' And

Montaigne agreed that it was a presumption for men to set themselves above the rest of creation. Their general duty was to be humane, 'not only to such animals as possess life, but even to trees and plants.'[6]

Such rare Renaissance sensibility counterpointed the arrogant dismissal of other species in the sixteenth century. Moreover, the nature and souls of the wild Indians of the Americas led to a confrontation and a searching of conscience in Valladolid. To crusade against the Muslim or to expel the Jew could be clearly excused by the Catholic faith; but to exterminate and enslave the American Indian needed the flexible logic of authority as well as the iron of the sword and the chain. The Spaniards had not conquered the New World for gold and land grants alone; the friars had followed the flag. And one of them, the Dominican Bartolomé Las Casas, wished to complete nearly fifty years' work for the Indians by forcing the crown to accept his policy for the peaceful conversion of the Indians to the Catholic faith. His method was to write a series of books in their defence, to lobby at Madrid for the passing of new laws to protect the rights of the Indians, and to force the public dispute at Valladolid against the leading defender of extreme colonial practice, Juan Ginés de Sepúlveda, an authority on Aristotle, who believed with his Greek master that slavery was the natural state of some men and that the world was split between city people and barbarians.[7] In his dispute with Sepúlveda, Las Casas was merciless in his search for mercy, a bigot in pursuit of charity, hungry for the sake of humanity. If he won his argument and lost his purpose, yet the gap between the laws of Spain and the facts of the colonies was as wide as the Atlantic, and as full of storms.

The trouble about the dispute between Las Casas and Sepúlveda was that it was held too late. Justification follows the crime. The 'huge and traumatic step backwards' of reviving slavery in Europe and its conquests was already a fact.[8] The Arabs of Granada had taught the Spanish warlords who displaced them that using slaves was natural. The Portuguese followed the Muslim example in Africa, and the Spaniards in the New World. Slavery was already accomplished by the time that Aristotle was called in to excuse the treatment. Although the conquistadors often had the grace to wonder at some of the laws and achievements of the advanced urban civilisations of the Aztecs and Mayas and Incas, yet they could justify their massacre and looting and debasement of the Indians by condemning them like Cortes as 'barbarians lacking in reason, and in knowledge of God, and in communications with other nations.'[9]

By some alchemy of the New World, every Spanish adventurer from some peasant village in Castile was transformed into a *caballero* by the mere fact of setting foot in the Americas, so that he became unfit to toil with his hands except in the art of administration or war. Like the medieval knight, his loot was his right. Like the Greek aristocrat, his ideal was leisure for the cultivation of his soul and his lands, although that leisure depended on the economic slavery of the inferior. Unfortunately, feudalism in Spain and Portugal had made the ideal of all men, however poor, that of the warrior lord supported by a peasant or serf population. If this serf population was barbarian and black or brown, if it could be enslaved to work a hacienda or a plantation, so much the better.

From the beginning, the Dominicans and the Papacy had protested against the maltreatment of the Indians and had tried to protect the remnants of the first Americans. In 1537, Pope Paul III issued the bull, *Sublimus Deus*, saying that the Indians were not to be treated as dumb brutes created for the service of Catholics, but as men and potential converts; even before conversion, they were not to be despoiled or made into slaves. In the long debate at Valladolid thirteen years later, Las Casas appealed for the protection of the American Indian by reason of experience and conscience, while Sepúlveda consigned the Indian to repression and exploitation by reason of Aristotle and the Christian Fathers.

Like many philosophers, Sepúlveda had stayed at home to think and knew next to nothing of the Indian; he based his case on the authorities of the past. Aquinas had said that certain wars were just. The Indians were sinful because they were idolators, and therefore war against them was just. Resistance by the Indians was, however, unjust; for they fought the soldiers of God. The Indians were barbarians, therefore they should be enslaved by the superior Spaniards in order to learn religion and civilisation. This would protect the weak Indians against oppression by the strong Indians, and it would bring peace and law where there had been savagery and anarchy.

Unfortunately, Sepúlveda chose the behaviour of the Spanish troops at the sack of Rome in 1527 to prove how virtuous they were; although they had burned and killed and looted and raped in the most holy city in Christendom along with the rest of the army, they had all repented on their deathbeds and had restored their loot. Perhaps Sepúlveda was claiming that Spanish destruction was followed by the virtues of Spanish rule, which made the destruction merely the necessary condition before salvation. What worth was all the gold and silver plundered from the Americas, Sepúlveda asked, against the introduction of iron to the continents by the Spaniards, who had also brought in wheat, barley, oil, horses, mules, oxen, sheep, goats, and many trees, not to mention writing, books, culture, laws and Christianity? The Indians were *homunculi* or little men, inferior to the Spaniards 'as children are to adults, as women are to men.'[10] In obedience lay their only way to heaven.

Las Casas, who had attacked Aristotle in his youth as a 'gentile burning in Hell', now decided to adapt the Greek philosopher to defend the American Indian, in the same way that Aquinas had adapted Aristotle to make a new foundation for Catholicism. If Aristotle in the *Politics* declared that the natural slave could not know virtue nor share in happiness nor in free choice, yet in his *Ethics* he stated that his idea of the slave did not imply inferiority or inequality because of race or status. Moreover, Aristotle always praised the virtues of diversity. Therefore, could not a good Indian become a good citizen in a different way to a good Spaniard?

In his huge *Apologetic History*, based on twenty years of personal observation of the American tribes, Las Casas set out to prove that the Indians compared well with the Greeks and the Romans, were certainly rational beings, and conformed to much of natural law. They were more religious than the ancients, because they gave better sacrifices to their gods; their education of their children and marriage arrangements were admirable; and the fact that Indian women worked with their hands for their families made them more virtuous than many Spanish women. Las Casas may have repeated Sepúlveda's error over the sack of Rome when he praised even the human sacrifices of the Aztecs as proof of their religious nature; but his catalogue of the virtues and inventions and civilisation of the Indians was a strong argument. The Spanish crown became slowly convinced by the supporters of Las Casas, and its laws for the treatment of the Indian in the Americas became steadily more humanitarian.

Yet colonists have never taken kindly to laws passed in the mother country. They presume that the home government is ignorant of actual conditions. Thus the ferocity of the conquistadors did not abate. In the same year that Las Casas was presenting his case against Sepúlveda, Pedro de Valdivia was cutting off the hands and noses of two hundred Araucanian Indians captured in a battle in Chile, because they had refused to submit to the Spanish crown. Valdivia excused the action by writing to the king that he thought this was the best policy 'for fulfilling Your Majesty's commands and the satisfaction of Your royal conscience.'[11] The man on the spot had, unfortunately, the power to act for the man on the throne. However well-meaning the arguments of the monks and the actions of the King's Council in Spain, their power to affect the immediate policy of the colonists in the New World was limited. The first viceroy in Peru had his head cut off by the colonists and bounced about on a string for trying to protect the Indians, and as late as 1599, Oñate was lopping one foot off all the rebellious male Indians from the pueblo of Acoma and sentencing them to twenty years of hobbling slavery. Even representatives of royal power were circumspect in the Americas for fear of provoking the colonists into rebellion against the motherland. As one Franciscan professor at the University of Mexico commented ironically, despite the fact that there were about four hundred verbal defenders for every Indian, the Indians continued to be treated as slaves or cattle.

Nothing showed the futility of Spanish legalism more than the reading of the *Requerimiento* before Spanish troops could rightly attack other peoples. The document was drawn up shortly before Cortes's assault on the Indian empires of the Americas, and every conquistador was required to carry the wording, in order to solve his ethical problems. The author of the *Requerimiento* was a believer in the just war and in Christian civilisation. The barbaric audience was ordered to submit to the authority of the Catholic Church, the pope, and the Spanish king under penalty of treason and expropriation. Refusal to submit made the Indians guilty of provoking the Spanish attack. The fact that the *Requerimiento* was read in an alien language at a distance that made it inaudible was no excuse for precivilised peoples. Whether they heard or not, they should yield and were responsible if they disobeyed. The *Requerimiento* put these words in the mouth of the conquistador: 'You shall be blamed for all deaths and losses, not the king nor I nor my soldiers.'[12]

This justification before mass murder merely proved what primitive peoples always discovered – that written laws are often an arbitrary collection of sentences for the convenience of the victors. Queen Isabella had asked what was the use of the first Spanish grammar, the first in any European language. The Bishop of Avila had replied, 'Your Majesty, language is the perfect instrument of empire.' So the Indians of the Americas discovered. The reading of the *Requerimiento* before a massacre did not make the punishment fit the crime, but created the crime to make the punishment fitting. For a scrupulous conqueror, a slaughter became a military and even a godly action, if the rules were observed and humanity was ignored.

The colonial vision of the Indian was often an inherited aversion to beings who lived in forests or wilderness. There was an instinctive fear of men and beasts which lived harshly on flesh and in ambush. As Verrazzano noted in his exploration of the North American coastline, the woods were so thick that 'any army were it never so great might have hid itself therein.'[13] The local Indians used trees in their defence, as John Hawkins discovered. 'They use a marvellous policy for their own safeguard, which is by clasping a tree in their arms and yet shooting notwithstanding; this policy they used with the Frenchmen in their fight, whereby it appeareth that they are people of some policy.'[14] The Frenchmen who set up the first colonial fortress in Florida were massacred by a Spanish force, that was in its turn exterminated by the French and their Indian allies. The reason for the disasters was surprise, because the woods stretched up to the walls of the fortress and the defenders had no warning of the coming of their foes. The French, indeed, hanged the Spanish 'on the boughs of the same trees, whereon the French hung.'[15] Those who cleared a patch in the forest and trusted in log walls for their defence lived in daily fear of sudden surprise and total destruction by fire or arrow or cord.

Thus the difficulty of taming the forest was complemented by the fear of attack from the forest. Since ancient times, indeed, the worst fate that could happen to civilised man was to be driven out of the city walls to live in the wilderness. Aristotle had argued that ostracism was the most necessary punishment of the Greek city state. The superior and extreme man must be excluded, for he was no longer a political and social animal like the rest. The Elizabethans could certainly understand the Greeks on that conclusion. In *Timon of Athens*, Shakespeare made his ruined and misanthropic hero turn his back on his city with the words:

> Timon will to the woods, where he shall find
> The unkindest beast more kinder than mankind ...
> And grant, as Timon grows, his hate may grow
> To the whole race of mankind, high and low![16]

There was no place in the small urban societies for the anti-social man. Even among the barbaric ancestors of the Elizabethans, to be sent away from the king's hall was the worst of fates. As the Old English lament, *The Wanderer*, declared, the life of the outcast was as the swallow. To stay inside the blood group was survival; to be excluded was death.

The first recorded encounters of the savage Americans of the ice and the woods with the barbaric northerners of Europe were prophetic. The way was already opened to the debate at Valladolid. Two Icelandic sagas, the *Graenlendinga Saga* of the twelfth century and *Eirik's Saga* of the thirteenth, told of the discovery of the North American continent by Norsemen and of their attempts to trade and colonise there.[17] The Norsemen had driven themselves across the Atlantic because of hunger for loot and land; sometimes they had been driven willy-nilly because their lumpish cargo boats could not hold their course in a cross wind with their single sail; thus they could land hundreds of miles off course at the end of a gale. The islands of the Atlantic were their stepping-stones, first the Orkneys, then on to Iceland and to Greenland, the base for expeditions to Helluland (probably Baffin Island), to Markland (probably Labrador), and then on to Vinland, where the explorers found the wild grapes of New England that do not grow north of Passamaquoddy Bay in Maine. There the western push of the Vikings faltered for fear of Indian attack and they began the long retreat across the Atlantic, until the last Norse colony in Greenland died out at the end of the fifteenth century, subjugated by the Eskimos who were better adapted to survive on the indifferent and increasing ice.

The Norsemen lumped Eskimo and Red Indian together under the sneering name of Skraeling, which meant 'wretch' more than 'savage', but was a word used towards despised inferiors. They were themselves a savage people; their ferocity in war had terrorised the coasts of Europe from the Baltic to the Bosphorus. Little better than pirates in the beginning, they produced a feudal caste of warrior nobles, who took over much of the coastline of Europe, also a hardy group of farmers and traders, who set up small and independent republics on the empty islands of the Atlantic. The histories of the Norse settlements in Iceland and Greenland show a tribal society emerging into a feudal one, with an aristocratic government gradually learning to control the unending violence of blood feud, and a pagan civilisation giving way to a Christian one. Life was bleak, barren and brutal on the stony farms; men and women were almost as unyielding as the soil and the winter; each existence was barely won and hardly lost. Blood kinship was the only certainty in life, and it was defended in blood. Each crossing of the sea might well result in death, and those who survived were lucky. The name, indeed, given to the supposed discoverer of Vinland in 1000 AD was Leif the Lucky, the son of Eirik the Red, although his brother Thorvald on a later voyage was the first to meet the Skraelings of North America and to import the violence of Europe into the American wilderness.

In the *Graenlendinga Saga* Thorvald was made to land in Vinland with his companions; he praised the beauty of the place and said that he wished to make his home there. The saga continued: 'On their way back to the ship they noticed three humps on the sandy beach just in from the headland. When they went closer they found that these were three skin-boats, with three men under each of them. Thorvald and his men divided forces and captured all of them except one, who escaped in his boat. They killed the other eight ... ' This slaughter provoked an assault by the Skraelings in their skin boats – certain of the Red Indian tribes of New England used canoes made of moose-hide rather than sewn birch-bark and they slept beneath their boats for shelter, as Jacques Cartier noticed during the first French expeditions to Canada.[18] The assault was repelled, but Thorvald was killed by an Indian arrow and his men sailed home to Greenland with their profitable cargo of grapes and vines.

The following expedition to Vinland was a colonising and trading one, led by Thorfinn Karlsefni, who was married to the widow of another of Leif Eriksson's dead brothers. He imported sixty men and five women into Vinland, with cattle and supplies, and they put a wooden stockade round Leif's log cabins. The Skraelings were at first terrified by the bull's bellowing, but in the end they began to trade. They were swindled of their furs and sables, getting in exchange milk and strips of red cloth. Karlsefni refused to allow his men to trade any of their iron weapons with the Skraelings for furs, and eventually a Viking killed one of the Skraelings for trying to steal a weapon. A general battle developed in which the superior weapons of the Vikings triumphed. Even when the Skraelings captured an axe, one Indian only used it to kill another and the dangerous weapon was thrown into the water. Karlsefni then left with his profitable cargo of furs and grapes. The Skraelings had no need to scare off the final expedition to Vinland under two brothers called Helgi and Finnbogi, for a sister of Leif called Freydis managed to get one half of the Vikings to kill off the other half so that she could get more of the profits on their return to Greenland. Karlsefni sailed back to Norway with the most valuable cargo ever to come from Greenland, and he even managed to sell his figurehead, carved out of the maple wood of Vinland, to a Saxon for half a mark of gold.

The *Graenlendinga Saga* told the classic tale of the encounter of European and Indian. Because the Indians were few, the Europeans were easily able to establish themselves in a base camp and to gather in supplies; the novelty of the appearance of the white men and of their cattle and foods allowed the Indians to be easily terrified and exploited; mutual fear of each other drove the European and the Indian to acts of aggression; the superior weapons of the European triumphed in the battle. It was the story of a slightly more advanced civilisation meeting a more primitive one. The Indian village was a mobile one of tents which Thorvald sighted in the distance, while the Norse settlement soon became a wooden village behind a stockade. The Indians used river canoes of hide, the Norsemen sailed ocean-going ships of timber. The Indians were trappers and hunters, although they were skilled in the use of the profuse vegetation of Vinland; the Norsemen were traders and farmers, prepared to import cattle in order to introduce their European way of life to this new continent. Above all, the Indians were a Neolithic people who fought in a group under a tribal leader; their weapons could not compete with the iron swords of the Vikings, who could fight each man on his own. Although the Indians were better adapted to survive in the wilderness of Vinland, the invaders exploited the natural resources of the continent remarkably quickly and set up their own way of life in a season. And when they returned to Greenland, they came back with a profitable cargo, making Vinland seem an Eden to the farmers of the barren north.

The later *Eirik's Saga* added much sociological detail about Karlsefni's expedition. There was a description of the Skraelings, who approached the Norsemen waving rattle-sticks to frighten the white men, or to exorcise them. 'They were small and evil-looking, and their hair was coarse; they had large eyes and broad cheekbones.' The Indians reconnoitred, traded, and then attacked. They used a catapult – a weapon of war hallowed in the traditions of the Algonquin Indians. The catapult terrified the Norsemen, who retreated losing two men. The Skraelings might have overwhelmed the Vikings, if the pregnant sister of Leif Eriksson had not snatched up the sword of a dead man, pulled one of her breasts out of her bodice, slapped it with the sword, and sent the Skraelings scampering away in terror at such a monstrosity. Again, the story of the dead man's iron axe was told. This time, a Skraeling broke it on a rock and threw it away, thinking it worthless because it could not withstand stone. So the Stone Age stuck to its ways faced with the Iron Age, even though the Vikings had set up a rudimentary ironworks in Newfoundland. The Indians won a stay of tenure on their continent. For there were enough of them to deter the Vikings, whose ferocious habits still recoiled at the fear of Skraeling savagery, and whose numbers were too small to settle a colony.

These early accounts of the Norse expeditions to America might mix fact with poetic licence – in one version, Thorvald was made to die from an arrow shot by a mythical Uniped. Yet the pattern of European reaction to the American wilderness and its inhabitants was established – the fear of the medieval monster and the savage leading to conflict, the exploitation of the Indian, the wonder at the riches of America and the quick exploitation of those riches, and the legend of Vinland seeping back to Europe to seduce later explorers and immigrants with the promise of abundance on the far side of the Atlantic. What did the New World seem to be, indeed, but a strangely familiar version of the Old, with virgin soil to replace overworked strips, with tall timber for tree stumps, with river estuaries for fjords, and only the fear of the violence of the Atlantic and the wilderness and the Skraeling to deter the coloniser?

Yet the last Norse colony in Greenland was dying out as Columbus sailed further south to another encounter of the peoples of America with Europeans. With him, he brought the ideas of the Old World to draw the lines of the New. Not only could the cartographers now begin to map the shores of the western islands and continent, but the philosophers and priests could describe native societies in terms of a golden age and heaven and hell. The letter which Columbus sent back to Spain about the New World made it seem an earthly paradise, nearer to dream than discovery. Among the nightingales and green fields and metal mines, the natives were naked and innocent. 'They are so guileless and so generous with all they possess, that no one would believe it who has not seen it.'[19] They were, as the court historian of Spain, Antonio de Herrera, was later to comment, 'a people in their original simplicity.'[20]

From the safe hindsight of the Age of Reason, Herrera wrote an account of the voyages and government of Columbus in the West Indies. He used the official archives and particularly the work of Las Casas. Unconsciously, Herrera described in his first volume the distance between the wonder of first encounter and the disillusion of colonisation. He began by carefully defining the other hemisphere as inferior, since it was less well adapted to human life because of its extremes of heat and cold. Failing to mention the benefits of the New World such as maize and potatoes, Herrera followed Sepúlveda in accusing the original Americans of having no European domestic animals, few European fruits, no wheat or rice, and, above all, no knowledge of iron and little of fire. 'They knew nothing of fire-arms, printing, or learning. Their navigation extended not beyond their sight. Their government and politics were barbarous.' Except on a few fertile plains and in the city of Cuzco in Peru, the people lived like savages. 'And as tame creatures are more numerous than the wild, and those that live in companies more than the solitary; so the people that live neighbourly in towns, and cities, are more political than those that dwell like beasts in the woods and mountains.'

After this definition of the Indian as barbarian, Herrera forgot himself enough to describe the natives of Hispaniola as good by nature and capable of Christian action. When one of Columbus's ships was sunk on a search for gold, 'the Indians so affectionately gave their help in this distress, that it could not have been better done in Spain, for the people were gentle, and loving, and their language was easy to be pronounced, and learnt; and though they went naked, they had some commendable customs...they would know a reason for every thing. They knelt down at the time of the *Ave Maria*, as the Spaniards did ...'

Nakedness was variously both the mark of innocence and of evil in medieval and Renaissance art. Sinners were stripped for hell fire, yet Adam and Eve also went naked in the Garden of Eden, while the cherubim and the saved flew bare to heaven before they were clothed in white. Nakedness meant only an absence of civilisation, and Herrera followed Columbus in presuming that the first natives met in the West Indies were in a state of nature before civilisation, a state which could certainly be influenced by Christianity for the good. Thus Herrera excused the Papal Donation of the new World, except for Brazil, to the Spanish crown. 'This Donation differed very much from what is usually granted to other Princes, *because it was not in prejudice of any man*, and because their Catholic Majesties had acquired a just title by temporal power for the promulgation of the Gospel ...' By this reasoning, the Indian could not be dispossessed, although innocent, because his society was not advanced enough for him to qualify as a man, until his soul was saved and he had accepted the Catholic Church and the Spanish state. Through the need for religious government, Original Simplicity was to be transposed to Original Sin.

Columbus, indeed, on his second voyage to the West Indies, found that the gentle tribesmen could also be savages. He lost some of his men for a week in the thickness of the woods of one of the islands and a relief expedition came back with tales of the cannibal Caribs, who were civilised enough to live in villages. 'They saw many men's heads hung up, and baskets full of human bones, and the houses were good.' So the savage displaced the simple in the forests, and Columbus had to build a stockaded fort in Hispaniola to protect the first colony which he left behind him. He himself had always treated the Indians discreetly, but as soon as he returned to Spain, the colonists began 'going about in an insolent manner, to take what women and gold they pleased'. The result was predictable. The peaceable Indians attacked the Spaniards, who broke them with men on horses and twenty 'wolf-dogs, the which in regard that the Indians were naked from head to foot, made terrible havoc among them.' The Indians captured in this war were adjudged to be slaves and many were sent to Spain. Tribute was now demanded or else labour on land, which was granted to Spanish settlers. Friars were also sent out to preach the Gospel, while territory and titles were assigned for the upkeep of the Church. The failure of the friars immediately to convert the surviving Indians on Hispaniola led to a complete change in Herrera's attitude to the natives. He now described them as people who had concubines and were addicted to sodomy. 'It plainly appeared, that the Devil was entirely possessed of those people and led them blindfold into error, talking and showing himself to them in several shapes; and that they were naturally of a mean capacity, given to change, and incorrigible.'

So was Herrera given to change. Having originally defined the natives of America as inferior and barbarous, he was seduced by Columbus's description of a paradise at Hispaniola into believing that they were simple and corrigible. Once, however, they had resisted the authority of the Spanish state, Herrera accepted the Church's denunciation of them as children of the Devil, although Columbus had stated that they knew no creed and did not worship idols. The wonder of first meeting had given way to the need to justify government, and in the case of Hispaniola, the further need to explain both slavery and total extermination. By 1510, the first black slaves were being sent to the island to work the gold mines, 'the natives being looked upon as a weak people, and unfit for much labour.' Also a general licence was issued to make slaves of all Indians taken in war, which in point of fact meant all Indians who could be captured. Within twenty years of Columbus's first landing, the large Indian population of Hispaniola was reduced to some fourteen thousand people. By the end of the sixteenth century, it had disappeared altogether, displaced by black slaves and white masters.

So Hispaniola developed the pattern of the encounter between the primitive tribe and the European explorer. The explorer was prepared for savage trickery and attack, but he dreamed of hospitality and welcome. Yet if he were well-treated, he was bound to abuse his hosts. For even if he did not himself exploit trusting natives, he came as the agent of a king or the spy of a trading or colonising nation. In the national epic of Portugal, *The Lusiads* by Camoëns, the voyage of Vasco da Gama to India in 1497 was modelled on the *Aeneid*, and da Gama set out with the help and malice of the gods to found a second Roman Empire in the East. After surviving Muslim treachery at Mozambique and Mombasa, he and his men were received royally by the King of Malindi. To the king, da Gama's envoy complained: 'But what barbarous races there are to be found in the world today, what savagery and bad faith, that will deny to honest men not harbours merely, but the hospitality of the very desert sands! What evil purpose or disposition is it they suspect in us, that they should fear so small a band, and lay snares to destroy us?'[21]

That evil purpose da Gama knew. He came to trade, colonise and Christianise, and he called it a good purpose. He did not have the cheerful amorality of an Amerigo Vespucci, whose sense of mission seemed always to have been subordinate to his sense of opportunity. At least, when Vespucci was writing to Piero Soderini about his first encounter with the Indians of South America, he knew the advantages of his faith and the lack of theirs. 'They showed themselves very desirous of copulating with us Christians. While among these people we did not learn that they had any religion. They can be termed neither Moors nor Jews; and they are worse than heathen; because we did not see that they offered any sacrifice, nor yet did they have any house of prayer. I deem their manner of life to be Epicurean.'[22]

The joys of the New World had their consequences. Not only did the first Europeans exchange hawk-bells for plates of gold, but they also exchanged diseases. If they imported smallpox into the Americas, they exported syphilis. 'By having to do with the Indian women, they contracted a distemper, common enough among the natives, but altogether unknown to them, which occasioned them to break out in blotches all over their bodies of which many died, and others thinking it to be cured by changing air, returned into Spain and spread the distemper there.'[23] The pleasure of innocent promiscuity seemed as blighted as the wonder of first meeting. It could turn to a Puritan revenge, as when Balboa burned alive or fed to his dogs more than fifty of the homosexual slaves whom he captured on the Isthmus of Panama.

The ferocious part of savagery has always had its edge whetted by morality and religion. Sadism finds its excuse in justice as war reveals its mercilessness in crusade. On meeting the invaders from Europe, the American Indians found the conquistadors fiercer than themselves. For the Spaniards used trained beasts as well as cutting metals in their attacks. Even if the Aztec religion demanded human sacrifices, the Catholic religion justified the massacre of the infidel and the rebel. Both Indian and European discovered savagery in each other, but in the inevitable conflict of a culture of iron with Neolithic cultures, the sword was bound to sharpen itself on the stone.

The only metals worked in the Americas before the coming of Columbus were gold and silver and a little copper, and these were used mainly for ornament. Being soft metals, they were often beaten rather than fired. After the Indians first met the exploring Spaniards, they would trade any quantity of gold for what they called white metal, iron and tin. The initial effect of the Spaniards' cannon and armour and crossbows and swords was decisive against semi-naked tribes armed only with stone-tipped spears and wooden clubs and bows and arrows. In Alvarado's massacre of the sacred dancers in Tenochtitlan, the Aztec chroniclers recorded the terrible slashing of the iron. The Spaniards 'attacked the man who was drumming and cut off his arms. They then cut off his head, and it rolled across the floor.' They attacked all the dancers, 'stabbing them, spearing them, striking them with their swords. They attacked some of them from behind and these fell instantly to the ground with their entrails hanging out.' Others were beheaded or had their skulls split into pieces. The Spaniards 'struck others in the shoulders, and their arms were torn from their bodies. They wounded some in the thigh and some in the calf.' Others had their stomachs cut open. 'Some attempted to run away, but their intestines dragged as they ran; they seemed to tangle their feet in their own entrails. No matter how they tried to save themselves, they could find no escape.'

The Spaniards in their turn had to watch the sacrifice of their fellows by the obsidian knives of the Aztecs. Once fifty-three Spanish prisoners and four of their horses were cut open by the Aztec priests, who removed their smoking hearts and set up their heads on pikes facing the sun, the horses' heads below the men's. But in the pitched battles, obsidian was no match for iron and the quilted armour of the Spaniards could absorb arrows better than naked flesh. At first, indeed, the Aztecs had thought horse and rider were one like a Centaur. Later, they thought the horse was a giant stag, which snorted and bellowed and sweated and foamed with its iron bells ringing at its neck. The shock of the horses was like the force of a storm or an earthquake. 'They make a loud noise when they run; they make a great din, as if stones were raining on the earth. Then the ground is pitted and scarred where they set down their hooves. It opens wherever their hooves touch it.'[24]

Although iron vanquished the stone empires of Mexico and Peru, the use of the animals of Europe, the trained horse and the dog, caused the most terror in the American Indians. When Ponce de Leon was putting down a revolt in Puerto Rico, he had a fierce hound called Bezerrillo, which tore open the Indians and knew 'which of them were in War, and which in Peace, like a Man.'[25] The Puerto Ricans were more afraid of ten Spaniards with Bezerrillo than of one hundred without him, and he received his share of the spoils with the rest of the soldiers. Cieza de Leon once met a Portuguese 'who had the quarters of Indians hanging on a porch to feed his dogs with, as if they were wild beasts.'[26] And when the Aztec emperor Motecuhzoma's messengers first reported to him the armaments of the Spaniards under Cortes, they described the cannon, the iron weapons, the horses, the blond hair of the white men, their sweet bread, and finally their dogs, which were said to be enormous with burning yellow eyes, tireless and powerful and swift and fierce and spotted like an ocelot. This last description terrified Motecuhzoma and helped to make him the indecisive puppet, resigned to his fate, whom Cortes manipulated so successfully on his first advance on the Aztec capital. Then the dogs were sent running ahead of the Spanish column: 'They lifted their muzzles to the wind. They raced on before with saliva dripping from their jaws.'[27]

Ironically, it was the savagery of the trained beasts of Europe that broke the Indians as much as the technology of iron. The hooves of the cavalry and the teeth of the wolf hounds were as destructive and more terrible than the long-range death flying from cannon and musket and crossbow. The monsters of war were imported into the Americas, where the great city of Tenochtitlan was larger than any in Spain and boasted at least a quarter of a million inhabitants. As Bernal Diaz de Castillo reported, when Motecuhzoma first showed the Spaniards the sights of the white-stuccoed city from the top of the sacrificial pyramid in its centre, 'some of the soldiers among us who had been in many parts of the world, in Constantinople, and all over Italy, and in Rome, said that so large a market place and so full of people, and so well regulated and arranged, they had never beheld before.'[28] In Herrera's opinion, Tenochtitlan surpassed Venice as a city built on water and was twice the size of Milan.[29]

Yet the sacrificial pyramid itself was enough to confirm the righteousness of Christian conquest. Gómara, who was later the secretary of Cortes and his biographer, claimed that the skulls of 136,000 sacrificial victims were exposed near the great pyramid of Tenochtitlan, and that the entrance to the temple dedicated to Quetzalcoatl – the white bird-god whom Motecuhzoma confused with Cortes – was like the mouth of hell. The door of the temple was carved 'in the form of a serpent's mouth, diabolically painted, with fangs and teeth exposed, which frightened those who entered, especially with the Christians'. Every chapel inside the temple was crusted with blood and stank with human sacrifice. The worship of snake and jaguar gods seemed to be the worst of idolatry and a confusion of bestial with human nature. The main duty of the Aztec warriors was to capture victims for the knives of the priests. They believed that the source of life, the sun, needed regular infusions of human blood to renew itself. With this horror as the root faith of the Aztecs, it was not difficult for Gómara to support Sepúlveda against Las Casas and to declare: 'It is war and warriors that really persuade the Indians to give up their idols, their bestial rites, and their abominable bloody sacrifices and the eating of men, which is directly contrary to the law of God and nature.'[30]

So the clash of European and Indian confirmed the worst fears of either culture. The ruthless slaughter by Alvarado of the sacred dancers seemed as shocking to the Aztecs as the human sacrifices to the war-god Huitzilopochtli seemed to the Spaniards. Both peoples were treacherous and deceitful in war, both tortured prisoners by flaying or the rack if they needed information, both used massacre and terror as instruments of policy and strategy. Both cultures were urbanised and organised and semi-educated, but the collapse of the Aztec Empire, soon followed by the collapse of the Incas, before a few hundred Spaniards can only be explained in terms of the sense of Christian mission of the more moderate conquistadors, such as Columbus and Cortes, and the ultimate ferocity and avarice of such mass-killers and plunderers as Alvarado and Pizarro.

The Aztec chroniclers, indeed, mocked the greed of the European, which they found inhuman. When Motecuhzoma first sent gold in quantity to the Spaniards, they were seen to be transported with joy, fingering the gold like monkeys. 'Their bodies swelled with greed, and their hunger was ravenous; they hungered like pigs for that gold.' When, also, the Spaniards searched Motecuhzoma's personal treasure-house, the Aztecs saw that the Spaniards 'grinned like little beasts and patted each other with delight.'[31] Later, when Pizarro seized the Inca Atahualpu and made him fill a large room with gold and silver to the height of his reach, the Inca did not gain his liberty with his ransom, but his death. The great object of the Spanish expeditions to the New World was gold, and it is remarkable that their success should have been so complete.[32]

To them that know what they have not, shall it be given. From them that know not what they have, shall it be taken. That is a law of the conquest of the savage.

3
THE SAVAGE WITHIN

THE ENGLISH AND French and Dutch found no El Dorado in North America, no urban empires on the scale of a European nation. As the scientist and mathematician Thomas Harriot related of the first English colony in Virginia, many of the settlers returned to England disappointed 'because there were not to be found any English cities', so the country seemed miserable to them. Yet he and John White, who was the first Englishman to paint and draw the Indians, sent back the long account of the local tribes that is the original social anthropology of Virginia. Harriot recorded the advantage in Indian war of European discipline, weapons and cannon, but he found that scientific instruments were the cause of the Indians treating the Europeans as divine. 'Most things they saw with us, as mathematical instruments, sea compasses, the virtue of the loadstone in drawing iron, a perspective glass, whereby was showed many strange sights, burning glasses, wildfire works, guns, books (writing and reading), spring clocks that seem to go of themselves, and many other things we had, were so strange unto them and so far exceeded their capacities to comprehend the reason and means how they should be made and done, that they thought they were rather the works of gods than of men ...'[1] So the technology of war first awed and destroyed the Indians, to be followed by the technology of science and the mechanisms of colonial control. The death of the armoured invader might deny his divinity, but no Indians could work the devices by which he navigated and ruled.

The original Virginian experience, when Roanoke was abandoned through want and Indian war, made Richard Hakluyt draft a set of *Instructions* in 1606 for the new colonists, which showed that Europeans had also learned from their encounters with the Indians. Hakluyt did not refer to the Indians as 'savages', but as 'naturals'. He advised the new settlers to avoid giving offence until they had planted and reaped their first crop of corn, and had set up their defences far from wooded country. If they shot at Indians, they must always hit them. And should any of the Europeans be killed or taken sick, this must be disguised. If the naturals saw the settlers were 'common men and that with the loss of many of theirs they diminish any part of yours, they will make many adventures upon you.'[2]

These *Instructions* were actually useless, for the Indians of Virginia already knew what to expect from the colonists. Thus they attacked the settlers on the first night that the ships reached Chesapeake Bay – 'savages, creeping upon all four from the hills like bears, with their bows in their mouths.' And yet, these savages did not attack when the surviving colonists were dying of fever and famine in their little fort, for 'it pleased God after a while to send those people which were our mortal enemies to relieve us with victuals, as bread, corn, fish, and flesh in great plenty, which was the setting-up of our feeble men, otherwise we had all perished.'[3] And so the aboriginal Virginians progressed from the neutrality of being 'naturals' in the *Instructions* to being 'savages' when they defended their land and 'people' when they brought help to their enemies. Thomas Hobbes, who was to believe in the brutish theory of the Indians, asserted in his *Leviathan* that truth consisted of the right ordering of names in men's affirmations. For the natives of Virginia, the various names of 'natural' or 'savage' or 'person' were reflections of their behaviour to Europeans, which did not prevent the truth of their displacement.

The native Americans had four roles for the early colonists – enemies to be resisted and evicted, porters and miners to carry loads upon and below the earth, potential souls to be saved for Christianity, and finally, their most useful role as the interpreters of the soil. If the Indians were savages, naturals and people in their first three roles as warriors and carriers and converts, they were also ordinary hunters and farmers to the untutored colonists. At Roanoke the English failed to grow sugar and oranges and lemons and wheat. So the settlers were forced to depend on Indian corn. They learnt the local method of cultivation. As Harriot noted, the Indians used a slash-and-burn method to clear the soil of weeds and fertilize it with ash. Then they made a hole for the maize 'with a pecker, wherein they put four grains with that care they touch not one another (about an inch asunder) and cover them with the mould again; and so throughout the whole plot.' He tallied up the uses of grass silk, worm silk, flax, hemp, alum, pitch, tar, sassafras, sweet gums, dyes such as sumac, okra, peas, gourds, herbs and tobacco.[4] The medicinal function of the herb was explained by the Indians, as well as methods of extracting flour from chestnuts. Without such disinterested demonstrations of how to live off the land, the first Europeans could hardly have fed themselves. And with the discovery of the potato, parts of Europe such as Ireland came to depend on the roots of America for survival as well as for luxuries such as chocolate, coffee and vanilla.

Yet those American colonists who were to live off Indian knowledge of the land were often to claim that the savagery of the tribes came from the land as well. The confusion between the wilderness and the nature of those who lived in it was always in the European mind. Verrazzano, for instance, found the Indians of New York and Connecticut with their oaks and cypress trees, 'very pitiful and charitable towards their neighbours' – the fertile soil and abundance of game made them hospitable and friendly. But as he progressed past Massachusetts to the north, and as the coldness and firs grew, so the Indians changed their character. 'The people differ much from the other, and seeing how much the former seemed to be courteous and gentle, so much were these full of rudeness and ill manners, and so barbarous that by no signs that ever we could make, we could have any kind of traffic with them.'[5] Geography and climate changed human nature. The savage man was the product of savage country.

This impression fearfully influenced the Pilgrims, when they debated whether to leave Holland for the unknown New World. William Bradford, the historian of the Plymouth plantation, was present at the discussion. The Pilgrims were deterred by the possible casualties of the sea and the miseries of an unknown land, where they would be liable to famine and nakedness and the lack of everything Europe could provide. In fact, privation in America might turn them into castaways, forced to exist like the natives already there, who ranged up and down 'little otherwise than the wild beasts.' The nature of these people was said to be 'cruel, barbarous, and most treacherous.' They not only killed men, but delighted in tormenting their captives 'in the most bloody manner that may be.' The hearing of such cannibal tortures moved 'the very bowels of men to grate within them.'

Prejudiced by travellers' tales against the American Indians and fearful of their own inadequacy on the rim of the unknown continent, the Pilgrims landed in the winter of 1620 on Cape Cod. Bradford records the terrifying sight: 'Besides, what could they see but a hideous and desolate wilderness, full of wild beasts and wild men? And what multitudes there might be of them they knew not ... For, summer being done, all things stand upon them with a weatherbeaten face; and the whole country, full of woods and thickets, represented a wild and savage hue. If they looked behind them, there was the mighty ocean which they had passed, and was now as a main bar and gulf to separate them from all the civil parts of the world.'[6]

Cut off from civilisation, the Pilgrims peopled the woods with their fears. 'Wild' was their description of both the landscape and its creatures. As Michael Wigglesworth wrote in 1662, the forest was:

> A waste and howling wilderness,
> Where none inhabited
> But hellish fiends and brutish men
> That devils worshipped.[7]

If the Pilgrims had not felt themselves the elect and chosen instruments of God, they would have had few defences against the terrors of the thickets other than the wooden stockade which they soon erected around their first town at Plymouth, where Bradford himself became the second governor. The Pilgrims fought the wilderness by denying it. As soon as possible, they set up churches which would exclude the ungodly, and they established town governments which would repress temptations and punish sins. Although the Puritan faith believed in the family as the basis of the good life, the family was not strong enough to deal with the evil put in Adam after the Garden of Eden. 'As man's nature inclines him to be sociable, so the connate corruption in fallen man, disposeth him to evil society; and children early discover the naughtiness of their hearts ...'[8]

An exclusive and defensive religion was necessary to keep out wickedness from even the godly. Not only did Indians threaten, but also immigrants, who fouled the towns of Massachusetts. 'We have made an ill change,' Nathaniel Ward complained, 'even from the snare to the pit.'[9] Some might think old England was the home of depravity, but so was New England. For men carried their own corruption within them and strong government could be as much a cause of wrongdoing as a cure for it. By 1642, William Bradford had noted that even the most notorious sins of the Indians such as sodomy and buggery were practised within the Puritan towns. The reason why sin could not be escaped was that the Devil fought hardest against the godly, who then turned to repression. 'So wickedness being here more stopped by strict laws ... so as it cannot run in a common road of liberty as it would, and is inclined, it searches everywhere, and at last breaks out where it gets vent.'

The Puritan fought the Devil and the savage within himself, and he called the struggle conscience. Where a godly man gave way to what he considered lewdness, the twistings of his morality were more exigent than the needs of his flesh. In the most famous of all colonial marriages between the European and the Indian, John Rolfe had to excuse himself to the Governor of Virginia for wedding Pocahontas. Although this union between Indian princess and English gentleman was a perfectly reasonable compound of love and diplomacy and convenience, Rolfe felt bound to protest that he did not marry Pocahontas 'with the unbridled desire of carnal affection, but for the good of this plantation, for the honour of our country, for the glory of God, for my own salvation, and for the converting to the true knowledge of God and Jesus Christ an unbelieving creature ...' After years of wedlock, Rolfe wrote that he had become worried by his life with someone 'whose education hath been rude, her manners barbarous, her generation accursed, and so discrepant in all nurture from myself that oftentimes with fear and trembling I have ended my private controversy with this: "Surely these are wicked instigations, hatched by him who seeketh and delighteth in man's destructions".'

Rolfe probably played down the attraction Pocahontas had for him; but it is clear that close contact made him identify her with the Devil because of her different habits. Only his righteous justification of himself as a missionary and a civilising agent could make him reconcile his 'seared conscience' with his lust for the primitive princess who had 'entangled and enthralled him in so intricate a labyrinth.'[10]

This view of the Indian as the seductive agent of Satan was particularly strong in those of a godly temperament. Bradford showed that the worst fear of the Pilgrims was for the white ungodly Devil in their own ranks, who might become the leader of the Indians. The Satan of the Plymouth plantation was a certain Morton, who set up a maypole, brought in other European degenerates and Indian women, and held midsummer orgies as the Lord of Misrule. The spectacle of the European outdoing the Indian in lewdness and barbarism was bad enough for the Pilgrims; but Morton also threatened the whole existence of the colony by arming the Indians, so that they could hunt fur and game for his profit. In the new Virginian colony, the laws of 1619 had condemned any man who sold arms or gunpowder to the Indians as a traitor, who should be 'hanged as soon as the fact is proved, without all redemption.'[11] Thus when the Massachusetts Indians had killed a few of the settlers with Morton's firearms, Bradford could well ask princes and parliaments for 'some exemplary punishment upon some of these gain-thirsty murderers, for they deserve no better title, before their colonies in these parts be overthrown by these barbarous savages, thus armed with their own weapons by these evil instruments and traitors to their neighbours and country.'¹² Morton was deported to England, but similar men were landing on the shores of the continent, interested in making money rather than in founding a colony of the godly.

In Rolfe's description of his relationship with Pocahontas and in Bradford's account of the doings of Morton, some of the permutations of the European reaction to primitive tribes could be seen. The fear of the unknown led the settlers to presume savagery in Pocahontas and the Indians, who were called barbarous in practice and 'discrepant in all nurture.' There was a genuinely savage response to the advance of civilisation, both when the Indians killed the Europeans who were displacing them and when they resisted the assumption of Christian superiority – part of Rolfe's motive in seeking to evangelise Pocahontas seemed to be his wish to turn her from a wildcat into a submissive Christian spouse. The colonists also shrank from actual contact with savagery, a reaction which made Rolfe think of the Devil when he saw Indian habits. Finally, there was horror at the European who revealed himself to be a savage, for Bradford found Morton far more satanic than the Indians whom he led astray. Often more terrible than the savage outside the stockade of the settlement was the savage within the ribcage of the Puritan, and his sternness towards all dissenters was frequently no more than fear of his own nature. As William Carlos Williams stated in his essays on the maypole at Merrymount, Morton might have 'laid his hands, roughly perhaps but lovingly, upon the flesh of his Indian consorts', but the Puritans 'laid theirs with malice, with envy, insanely, not only upon him, but also – one thing leading to another – upon the unoffending Quakers. Trustless of humane experience, not knowing what to think, they went mad, lost all direction.'¹³

If the threat of the savage in the forest dismayed the Puritan, yet he came from a culture of timber that could use the trees for its houses and stockades and ships. His fear could be converted into his opportunity. The barrier before him provided the means for his progress. Once the descendants of the Puritans had broken through the backwoods on to the great plains – making the forest their familiar in the process and displacing the Indians of the trees in their advance – then indeed they lost all direction, faced with a new environment even more hostile than the timber which had once terrified them. The gigantic oval of the dry grasslands between the forests of the eastern seaboard and the Rocky Mountains was already the home of a great beast. There lived one animal 'that came nearer to dominating the life and shaping the institutions of a human race than any other in all the land, if not in the world – the buffalo.'¹⁴

This huge creature bred in its millions on the sparse grasses and became the main food of the mounted plains Indians who, for the two centuries after the Spaniards had brought them the horse, dominated their environment by their riding. Just as the Huns and Mongols of the steppe and the mailed knights of the open field had proved invincible, the horsed Indians proved the finest cavalry of the continental flatland. They followed the buffalo and their enemies in moveable villages, and they were held to be the wildest and the most ferocious enemies of the American settlers. When most other Indian tribes had turned to agriculture, the plains tribes remained nomads and hunters, pursuing their meat and fat across the grasses.

As the new American colonists displaced the forest Indians and drove them on to the prairies, some tribes adapted and some were decimated by the mounted Indian tribes. The Saukees of the woods were warned of their future fate. 'When hunger drives you from those woods, your bodies will be exposed to balls, to arrows, and to spears. You will only have time to discharge your guns, before, on horseback, their spears will spill your blood ... As you have seen the whirlwind break and scatter the trees of your woods, so will your warriors bend before them on horseback.'[15] Following the Saukees came the timber culture of the white pioneers, moving west, and they were also helpless on the plains outside their wheeled wooden waggons, which served as defence and shelter and transport for tools and goods.

Industrial civilisation in America began with the building of log cabins, but faltered on the edge of the prairie in the sod hut. There was little wood or water in this new hostile country, which the early Americans thought uninhabitable and called The Great American Desert.[16] It was two thousand miles of space to be passed through, not occupied. It was no place for the settler, only for the vagrant. An army captain was to observe in 1852 that the last strip of forest before the great plains, called the Cross-Timbers, seemed 'to have been designed as a natural barrier between civilised man and the savage.'[17]

The plains were not to be the final barrier. The Rocky Mountains and their foothills lay between the prairies and the Pacific coast. And these peaks, thickly wooded up to the height of the timberline above which the cold and the wind ensured that no tree grew, turned the pioneer American trappers into a bestial form of man. The 'Mountain Men' were the walking fears of the Puritans – anarchic savages dressed in skins and living on raw meat, without law or conscience, almost as fierce as the grizzly bears that disputed their hunting rights over the Rockies. To exist in that environment, the Mountain Men slipped backward in the scale of civilisation until their wilderness skills equalled those of the Indians. Moreover, hunting alone rather than in tribes, their reversion to the primitive could be at a lower level of savagery than the red men. One was called 'Cannibal Phil' because he survived the winter by eating his squaw. All of them ate 'anything that walked, swam, wriggled or crawled.'[18] Like beasts, they lived between famine and feast, drinking the blood of the buffalo and eating its liver raw, spiced with the gall from its bladder. The time of the annual sale of their beaver pelts was a time of drunkenness and orgy and killing and waste. The rest of the year was an animal life on the mountains that made a man forget that living was more than survival.

The beaver were overtrapped and the Mountain Men passed away, to be followed by the waggons of the overland pioneers, making to the west. The Rockies were to stop the Mormons, seeking Zion and religious freedom, and they in their turn were to help and prey on the later caravans lumbering through the passes towards California. When Indian attack or internal quarrels or lack of supplies did not cause the wheeled villages of the pioneers to break up into quarrelling groups which forgot the conventions of social life, a winter besieged by snow and want in the mountains could lead even Victorian women into a recognition of the savage within their niceties. The notorious Donner party lasted through the winter on the high sierras by consuming one another dead and alive. Its only concession to civilisation was to label the remains so that a man or a woman might not eat their own kin. The hardier pioneers lived by denying their taboos, and the horror of their fight for survival would haunt their successors on their way through the mountain passes.[19]

So each new barrier that confronted the immigrant became the receptacle of his fears. To the first European settlers, the ocean was the prime wilderness to be survived, and the fierceness of storms seemed to be reflected in the rapacity of the pirates who preyed upon sea traffic. Then to the settlers on the coastline, the forest with its beasts and its Indians was the frontier of terror. Once the forest itself had been cut and used and controlled, then the prairies became the new horror and their horsed Indians the most implacable and bestial of all men. And finally, there was the obstacle course of the mountains, which could turn Europeans themselves into beasts and cannibals before they might reach the west coast with its opportunity for a life of ease. The adjective 'savage' may originally have been used to describe the correspondence between woods and their wild inhabitants, but it came to mean any enemy of the mountain or of the ocean or of the 'inland sea' of the plains, whose ferocity and resistance seemed to be induced by the harsh nature around him. The savage being or place was truly an encounter and a frontier between oncoming civilisation and its destination. It was also the beast lurking within the labyrinth of each man's desires.

Ever since Plato, moralists had feared the animal in man's nature. In the *Republic*, Plato wrote of the danger of dreaming. 'Then the Wild Beast in us, full-fed with meat and drink, becomes rampant and shakes off sleep to go in quest of what will gratify its own instincts.' Evil thoughts or wants were described in terms of black horses or wild asses, pigs and hawks, lions and wolves.[20] From the beginning of European philosophy, men had suspected their wants and given to their appetites the fangs and claws of other species. As late as the age of psychiatry, one of Freud's more famous cases was called the Wolf Man. Of course, these transferences could only take place when human beings themselves were transferred from a natural place to an artificial city. This divorce created the brutish names for those carnal desires which they had brought within their walls from a wilder previous existence. For the savage and the beast outside the pale, however, there was no separation from nature. They were what they were within their familiar surroundings and themselves. Yet until modern times, they would have no recorded voice to defend their necessity and identity within the great chain of being. They would only be accused by the civilised of being who they were.

In medieval China, the barbarian and the beast without were tamed, the beast within was controlled by symbolism and society, and the whole of alien creation was excluded by a Great Wall and a refusal to cross the oceans. The Mongol conquest of China in the thirteenth century provided a brief dynasty. The sackers of Peking were changed into the consummate rulers of the ancient land. One of them, Kublai Khan, became legendary in Europe as the creator of an earthly paradise according to the descriptions of the Venetian traveller, Marco Polo. He was also the protector of the Silk Road that connected Asia to the Mediterranean through the Mongol domains. Paradoxically, the art of printing, the mariner's compass and firearms were imported into Europe in what was called the age of nomad peace. The destroyers of the steppes were the protectors of commercial intercourse and communications between distant cultures.

The fearsome invaders on their terrible horses did not replace the beasts of Chinese mythology, which represented ancient aspirations and fears. The dragon or serpent was still held to control the forces of the earth and the winds, while monkeys made minor gods of mischief. The telling of legends about their doings exorcised some of the animal instincts in man. And the policy of the Ming dynasty, which drove back the barbarians over the Great Wall, was to exclude the foreign and the uncivilised. Although Chinese junks were capable of competing with Arab dhows for overseas trade, command of the Pacific Ocean was finally left to Muslim traders. A great fleet sent to occupy Japan sank in a storm. And when European ships began to trade and colonise in the Far East, Ming policy was not competition, but control and retirement. Shipbuilding was abandoned, alien merchants restricted to concessions, the self-sufficiency of China declared a state policy. Japan also retreated into a splendid isolation.

This defensive pride was the opportunity of the seapowers of Europe, which were open to new influences and markets. The cannons of their ships were irresistible, and global commerce and gunpowder empires began to grow abroad.[21] With them travelled the parasites, to which the Europeans were becoming more immune, but which were fatal to distant tribes and islanders. The most significant consequence of these early contacts was biological. For instance, some five million Indians lived in the Americas at the arrival of Columbus; by 1900 the figure was to fall below four hundred thousand people. Smallpox and other germs did more to ensure white conquest than pioneers or gunboats. What was lethal for the natives of Asia and the Americas and Africa was the virus and the bacillus. These were the microscopic beasts within the Europeans, the invisible organisms which killed the peoples in the lands which the white men wanted.

Epidemic was even more powerful than gun and ship and steed in the economic mastery of the globe begun by the sea-powers of Europe. While fearing the savage, they brought with them virulent diseases, which would decimate the savage. When the Portuguese first sailed to the Congo, they found a people dependent on the palm-tree for oil and wine, for vinegar and a kind of bread, for fibre and clothing, for roofs and snares. They were welcome, until they introduced the illnesses, to which they were hardened, although they themselves suffered badly from exposure to malaria. They also commercialised the slave trade, which was customary in those parts. In the end, the introduction of the rubber tree from Asia would exploit and reduce the Congolese to the abject condition described in Conrad's *Heart of Darkness*. But that late biological imperialism was preceded by the parasites and economics of the northern seafarers, who never lost an opportunity to turn a forest culture or a barbaric people into a profit at an inhuman price. By the sixteenth century, plague and famine had swept the Congo, and one traveller found the situation so bad that 'forced by necessity, the father sold his son, and the brother his brother, everyone resorting to the most horrible crimes in order to obtain food.'[22] The people of the trees, indeed, were turned into savages.

4
THE SAVAGE ENSLAVED

CERTAIN GREAT CRIMES do stain the centuries. The transportation of some twenty-four million men and women and children from Africa across the Atlantic to serve as slaves was the greatest of the crimes of Europe. Nine million of the victims died on the voyage across, fifteen million survived to toil in the Americas. The very mass of the slaves involved in the trade degraded their individuality and their condition. As in the later case of the Jews in the Nazi concentration camps, the size of the operation reduced its victims to the status of animals. 'Markets of men are here kept,' a slaver wrote from West Africa, 'in the same manner as those of beasts with us.'[1]

So they were. The African captives were snared like game, physically examined as closely as horses, bought and branded like cattle, herded in a barracoon like pigs, chained below decks like wild beasts, then penned and led out to labour in the American fields under the whip like donkeys until they were worn to death. Slavery, as Voltaire said, may have been as ancient as war, and war as human nature. But no civilisation had ever subjected the ancient institution so stringently to the laws of commerce. The mechanisms of Europe, from the account-book to the design of the between-decks of the slave-ship, from the auction in Charleston to the drunken haggling with the kings of Bonny, was designed to degrade man's view of man. The Protestant slavers did not baptise as the Portuguese had. There was no mission now in Africa, only money to be made. The crews on the slave-ships were the refuse of the docks, for they were as likely to die of disease in the stinking holds as they were to be abandoned penniless and sick in the West Indies, themselves servants to a system that counted men's work only in ledgers. The economic basis of slavery corrupted whomsoever it touched and freed through wealth only some thousands of citizens in a few European cities and on plantations in the Americas.

For slavery in Africa was not a process for the accumulation of capital.

There was no local method of doing this. Among the Ashanti, for instance, the number of slaves possessed by a man represented his place in society, not his wealth. For slaves were allowed to keep their personal property and the hard-working slave enriched himself, not his master. Only in the European system was economic robbery added to personal slavery. Once the slave was considered as a mere unit of production, arguments could be held seriously in Jamaica about the return on capital if slaves were worked to death quickly and new imports had to be bought, as opposed to the profit when slaves were worked to death slowly and bred their own replacements. In investment terms, treating people as beasts made the calculus more simple.

Apologists for the slave-trade used repetitive arguments. The first was the rescue of the African from savagery. On one of the rare occasions that he opposed Dr Johnson, Boswell praised the slave-owners. He maintained that even if extreme cruelty was practised on the African savages, yet a portion of them were saved 'from massacre, or intolerable bondage in their own country' and introduced 'into a much happier state of life.' Of course, such a tarnished version of African village society and such a gilded version of American plantation existence were dependent on the biased reports brought back by the slave-traders themselves. For the civilised European needed to be able to justify this distasteful source of his wealth.

When an actual freed and literate slave, Job ben Solomon, appeared in Georgian London, he delighted the capital's society and did not change the slavers' habits. As Montesquieu had noticed in 1748, Europeans could not consider Africans as men, 'because, allowing them to be men, a suspicion would follow that we ourselves are not Christians.'[2] Thus Job ben Solomon had to be treated as an exception, an African prince. Dr Johnson might fulminate in his introduction to a collection of voyages of discovery: 'The Europeans have scarcely visited any coast, but to gratify avarice, and extend corruption; to arrogate dominion without right, and practise cruelty without incentive.'[3] For a mercantile power like England, however, the profits of the trade were its justification, especially as its critics had to depend on the field-work of the first anthropologists in Africa, such as Barbot and Bosman and Phillips and Snelgrave, who were slavers themselves and were bound to paint Africa as black as possible to whitewash their own actions. Accusations of incessant cannibalism, massacre and torture made Africa seem a hell from which its inhabitants might beg to be transported.

The second apology for enslaving Africans was based on their nature. Barbot was unequivocal about them. They were savages through geography and temperament, 'generally extremely sensual, knavish, revengeful, impudent, liars, impertinent, gluttonous ... deceitful in their dealings with the Europeans, and no less with their own neighbours, even to selling of one another for slaves, if they have an opportunity.' Moreover, they were 'so very lazy, that rather than work for their living, they will rob and commit murders on the highways, and in the woods and deserts.'[4] If Barbot did not ask himself what profit there was in trying to make plantation labourers out of such useless material, yet he made his search for profit seem charity to the African refugees. Snelgrave, who visited Africa at the end of a brutal victory by the King of Dahomey, gave a much more practical reason for slavery. What was the use, he asked, of sacrificing prisoners instead of selling them?[5]

For Africa was no Utopia when the Europeans reached its coasts. Old Calabar on the Niger Delta, for example, was rightly notorious as a slaving centre; Mungo Park found that slaves outnumbered free men there by three to one.[6] The local Efik people and the Ibibios were disliked by the rest of the Coast tribes. They had an indifference to human beings and a wish for death that made them ghastly to other West Africans. Their ruling secret society, called Egbo or Leopard, flogged and killed to keep order; its followers appeared with sword and whip, hidden behind a demoniac mask with long raffia hair. At the funerals of kings, wives were strangled and followers beheaded or buried alive. The journal of a slave-trader described the death of a local town chieftain in these words: 'So we got ready to cut heads off, and at five o'clock in the morning we began to cut slaves' heads off, fifty heads in that one day ... and there was play in every yard in town.'[7] With such contempt for life dominant in Old Calabar, the captains of the slave-ships could feel merciful as well as commercial. They were, after all, taking the blacks from what Mary Kingsley was to call 'the steady kill, kill, kill' of West Africa.[8]

Another argument by the apologists of slavery was the most truthful. The vested interests of the trade made it of interest to all. As Snelgrave wrote, its benefits outweighed its real or pretended mischiefs. 'It will be found, like all other earthly advantages, tempered with a mixture of good and evil.'[9] The European traders made money, the Africans escaped to regular work and possible Christianity, the American planters secured a labour force strong enough to cultivate sugar and tobacco and cotton and rice. After all, economics were paramount, and as the enslaved Indians had largely died off, labour had to be secured from somewhere. White convicts and servants were more trouble and less productive than black people, who were acclimatised to the heat of the tropics.

THE SAVAGE ENSLAVED

Recent research on the political economy of slavery strongly suggests that it did not raise the slaves from savagery to civilisation. Many of the slaves did not come from herding or forest cultures, but from farming or even town societies. The black people from West Africa were not Neolithic, but already worked in copper and tin, while using iron hoes. A plantation system based on slavery already existed in Dahomey among a people renowned for their industry. There is little truth in the apologist themes of American historians of the Deep South that slavery was a progressive institution which domesticated the savage rather as the wild turkey had been domesticated. While certain of the forest tribes of Africa were still at a primitive stage of civilisation, the people of Ashanti and Dahomey, the Fulanis and the Yorubas, already had complex systems of trade and government, agriculture and the division of labour before they met Muslim or European influence.

The slave system, however, enslaved the plantation owners of the Americas in their turn. They held to the myth that they were free, independent, aristocratic and not obsessed by the economic considerations of those who lived in industrial cities. In fact, the capitalist's dependence on his factory workers was hidden, while the owner's dependence on his slaves was naked. As the blacks were blatantly the gatherers of the crops and the wealth, their masters could not escape the fact that they lived on their slaves any more than they could ignore their misdeeds while surrounded by mulatto house servants. If lip-service was paid to the values of patriarchy and purity rather than economics and passion, it was because the closer the crime, the blinder the eye and the louder the denial. As Hesse once remarked, second only to memory in importance to mankind is the capacity to forget.

The slave system brought the savage inside the plantation house. Among the kidnapped Africans were witch-doctors. These obeah practitioners used spells to terrify slaves and master alike. Voodoo and poison could maim and kill any enemy, black or white. Fetish oaths were at the back of slave and colonial uprisings, from the Jamaican rebellion of 1760 to the Mau Mau in modern Kenya. The ancient sorceries of the forest were a form of infiltration of the civilities of the plantation house; they might become an insurrection to burn it down.

More insidious for the outnumbered plantation owners was the erosion of moral standards. When even the Pilgrims had found it hard to keep to their Puritanism on the edge of the backwoods, how could the slave-owners hold to their civilised standards with the African savage as the servant inside the house? The expectations of African polygamy and European monogamy were so different that the result was confused enough without the added powers of masters over slaves. 'God forgive us,' wrote Mrs Chesnut of South Carolina, 'but ours is a monstrous system, a wrong and an iniquity! Like the patriarchs of old, our men live all in one house with their wives and their concubines; and the mulattoes one sees in every family party resemble the white children. Any lady is ready to tell you who is the father of all the mulatto children in everybody's household but her own. Those, she seems to think, drop from the clouds.'[10]

So the habits of black and white confused and influenced each other through slavery. The black mammy reared the white boy, the white youth abused the black housemaid, his father worked black men and women for profit, and all insidiously penetrated each other with their attitudes. The stolen Africans became African field-hands on their way through generation after generation to the American city. But as they passed, they left behind them the influence of the savage, the rhythms and the songs, the recognition of the violence beneath the skin, of the contradiction between manners and motives. White and black were raised side by side, and as the one denied the other, so he had to accept him against the grain.

The continual rapes of slavery, which brought the forest and village peoples of Africa to the plantations of America, also brought out the beast in their new white masters, whose violence and sexual fear could make them practise barbarities inhuman even to Old Calabar. In Jamaica in the seventeenth century, rebellious blacks were nailed to the ground and burned slowly to death limb by limb up to the head. They would be castrated or flayed or slit open with pepper dropped into their wounds. In the southern States of the nineteenth century, the lynchings of black men were a commonplace in fearful times, the bodies burning alive as they hung screaming from a tree. Man's horror of the alien and his fear of its revolt unleashed the violence within him. His effort to master his fellow men for his gain and to treat them as beasts discovered the brute in himself. If Africans were sometimes rescued from savagery, they often discovered the savage in their masters.

In Africa itself, the few Europeans clustered on the rim of the unknown land mass, building their little fortresses from Senegal to the Congo. The Dutch even shipped out the stone doorways of Elmina castle to keep back the dark and fearful forests of the continent. Although the lure of the gold mines of the fabled Mansa Musa of Mali sometimes took explorers up the Senegal and the Gambia and the Niger and other great rivers that lead to the interior, the hostility and power of the inland tribes did not encourage settlement nor conquest. The middlemen of the Sahara trade, the Arabs of the desert and the African kings and chieftains of the bush and coast country were left undisturbed in their dominions. As they had no written records, only the accounts of a few Europeans remain to describe the shock of encounter and to reveal the prejudices of the civilised mind.

The first and most obvious prejudice was skin colour; this difference was held to be enough to elevate the meanest white man above the noblest black. The Christian religion had for so long associated the colour white with God and His angels and the colour black with Satan and his devils, that even the bronze Ethiopians in their holy images represented Christ and his Ethiopian disciples as white, while the Devil was as black as the southern Nubian in his swamp or jungle. The slavers to Africa made use of this judgement by colour to justify themselves and their brutal trade. Captain Phillips in 1693 was unique in raising up his voice against such an invidious distinction. 'I can't think there is any intrinsic value in one colour more than another,' he wrote, 'nor that white is better than black, only we think so because we are so, and are prone to judge favourably in our own case, as well as the blacks, who in odium of the colour, say, the Devil is white and so paint him.' [11]

In the forest, of course, the colour of shadow was dark, also the colour of night. This absence of light was taken by Western thinkers and missionaries to be analogy with an absence of enlightenment, so that a dark skin seemed to imply a dark mind. In Livingstone's later diaries of his travels in central Africa, he was to rejoice in the fact that the black tribes all associated beauty with fairness, while he sought the blessing of divine power to enlighten 'dark minds as these'. So ineradicable were the colour judgements of most Europeans until this century, that their moral world was indeed equated with a visual one of black and white, bad and good, slave and master.

The other misunderstanding between men of different colours was their habit of accusing a whole alien group of the crime of one of their members. If an American colonist were killed by a black slave or a red Indian, this was taken to prove that all black and red men were intrinsically bad. The single death of a white could be the justification for the massacre of a whole African or Indian tribe. A colour difference identified the particular with the whole, the individual error with a racial wickedness. Again, it was rare before the nineteenth century that American voices of reason were raised such as Benjamin Franklin's, who once protested: 'If an Indian injures me, does it follow that I may revenge that Injury on all Indians? It is well known that Indians are of different Tribes, Nations and Languages, as well as the White People. In Europe, if the French, who are White People, should injure the Dutch, are they to revenge it on the English, because they too are White People?'[12]

Another prejudice was to relate the African tribes too closely to their environment. With no prescience of Darwin, James Houston could declare in 1725 that the customs of the Africans of the Slave Coast exactly resembled 'their fellow creatures and natives, the monkeys', while their natural temper and government were barbarous, selfish, deceitful, and uncivil. [13] The early scientist, Edward Tyson, published a treatise in *The Anatomy of a Pygmie compared with that of a Monkey, an Ape, and a Man*. He concluded that the pygmy was not a member of the human species; but, then, he had been supplied with the skeleton of a chimpanzee. European disgust at the unknown had led to a biased judgement, which confused the monkeys of the forests with the men who lived near them, so that the savage seemed at once a locale, a brute and a determining influence on human beings.

From an opposite prejudice that uncivilised man was naturally good, Le Vaillant would praise the primitive Hottentots after his travels among them in the 1780s. He was to admit that the Hottentots all slept together promiscuously in the same hut, but declared that they respected each other, brother and sister, mother and son. To say as other commentators did that the Hottentots lived like brutes was 'to calumniate innocence, and offer an insult to nature.' A savage was neither a brute nor a barbarian. 'The real monster is he who sees crimes everywhere, because he supposes them; and who asserts their existence on the odious testimony of his own conscience.' It was an early expression of Wittgenstein's insight that he who selects another as his target does not shoot at his enemy. His target is himself.

In defence of the Hottentot, Le Vaillant contradicted Houston and predicted Darwin's theory of evolution. He declared that a modern wit or physiognomist would assign to the Hottentot, in the scale of beings, a place between a man and an orang-outang, but that the qualities within the Hottentot did not allow him to be so degraded. His figure, in Le Vaillant's eyes, seemed sufficiently beautiful because the Frenchman had experienced the goodness of his heart. The Hottentots, timid and hospitable and immediate and pastoral, incited the idea of mankind in a state of infancy before its corruption.

So in Africa, explorers found both the bestial and the innocent according to their prejudice of the savage. Yet their innate biases and observed conclusions did not affect the transportation and extermination of the primitive Africans. Like the Indians of Hispaniola, the trusting Hottentots were eliminated by the Boers and by the warlike northern black tribes for not being able to become slaves. Le Vaillant knew that the deserts of the Hottentots were their only protection. He advised them to destroy any gold if they ever discovered it, for fear of wishing a Pizarro upon themselves. He even understood their occasional counter-attacks upon their enemies, particularly the Europeans. For what, so far, had been the history of the civilised penetration of three-quarters of the world? 'In every place where we have thought proper to establish ourselves, we have compelled these unhappy wretches, persecuted to a state of slavery, to betake themselves to flight; we have appropriated to ourselves, without the least scruple, whatever we found useful to us; and when the hour of vengeance has been proclaimed for them, and when they measured their blows by the magnitude of their injuries, without reviewing our own conduct, and too much blinded by interest or fanaticism, we have dared to call them barbarians, eaters of men, and ferocious animals, who live by murder, and allay their thirst with blood.'[14]

So the first explorers of Africa imitated the first discoverers of America in finding the beast or the child in the wilderness; they also saw either a hell on earth or a lost Eden. Yet later accounts reversed the bias. The slave trade was no longer to be excused in Victorian times, but to be ended. The odium of the sale of human beings was transferred on to African chieftains and Arab slave-traders, while the Europeans looked for souls to convert to Christianity. A black picture of African customs was painted to provoke funds for missionary effort or government intervention to protect trade; but within that murky picture, the black man himself was depicted as malleable, redeemable by mission schools. 'No one can conceive the terrible depravity to which the children of Ham have sunk,' David Livingstone wrote in 1853 from among the Makololo in Central Africa, 'unless he has intimate knowledge of their language and customs, but to change their savage nature is the object of our mission and it will be accomplished by the power of God's spirit through the gospel.'[15]

Little sense of Christianity drove the other Victorian explorers to the centre of Africa. They sought the glory of the discovery of the sources of the Nile, which haunted the imagination of the middle of the nineteenth century as El Dorado had obsessed Columbus and Orellana and Raleigh. The journals of Speke and Baker were as robust and self-centred as their drive to the twin sources of the White and Blue Nile. They did not blaze the trail for an empire in a fit of absence of mind. They moved in total surety of self, and none so much as Sir Samuel Baker, a thorough man, who carried his prejudices against the savage wherever he went. Only such certainty could have had a sponge-bath borne around Central Africa for two years, along with a clean kilt, sporran and Glengarry bonnet, with which to overawe an African chieftain several thousand miles from home. Baker was forewarned and forearmed to his regular teeth; never was explorer better prepared with guns and trade goods. The savage begged for mercy when Baker raised his sights.

Imperturbably, Baker outwitted slave-traders, cataracts, sunstrokes, hippopotami, fever, and vermilion-coloured Africans twanging at him with poisoned arrows. He slaughtered every formidable fauna and noted or ate every fanciful flora. He was crass in action, yet acute in description; the caricature of the sporting gentleman, but a genius of competence. He had the answer for everything from defeating a swordsman by ramming his umbrella down his opponent's throat to using fishskins to bind rifle-butts. With fist, bullet or stiff upper lip, he outfaced every situation by refusing to recognise it. Only at the end of his wanderings did he notice that the young wife whom he had dragged with him now looked old, and even she was worth his condescension. She was, he said, 'no screamer'.

Yet Baker's judgements of Africa were brutal and patronising, prophetic of the imperialism to come. He was acute enough to know the mechanisms of future control. 'The savage must learn to want; he must learn to be ambitious; and to covet more than the mere animal necessities of food and drink.' But present influence on such degraded people must be based on naked strength. 'Nothing impresses savages so forcibly as the power to punish and reward.' As for their aspect, Baker considered that his pet monkey looked 'like a civilised being' compared with the Nuehr savages. There was no law in Central Africa except for brute force, and human life was of no value. Indeed, Baker suggested that the natives of Central Africa had existed before Adam, since the historical origin of man (or Adam) commenced with a knowledge of God. These savages were like fossils, creatures roughly in the shape of humans, but lost from history. Their nature was on the level with the brute and 'not to be compared with the noble character of a dog.'

On one matter, however, Baker conceded that the nature of the African was deformed by nature itself. Given the appalling conditions of African life, broken by interminable massacre and slavery and treachery, the natives behaved better than many Europeans would have done in the circumstances. They had to be forced to witness public executions, which European ladies flocked to see. Once, although only once, Baker even saw himself ironically through their eyes. On reaching the lake which he christened Albert N'yanza, he ordered his men to give three cheers for England and the sources of the Nile. 'Hurrah!' they had to shout. 'Hurrah! Hurrah!' Then Baker observed that it was a 'wild and to their ears savage, English yell.'

Such philistinism in the jungle preserved the English explorer in his cocoon of contempt. The savage was despised, not feared; the alien was circumvented, not admitted. Baker confessed to enjoying 'a real difficulty', such as not being allowed to proceed. He merely went on, ignoring everything that did not suit his purpose or his righteousness. The measure of the value of crassness is that he was far more successful in Central Africa than his contemporary Sir Richard Burton, the great scholar and linguist, subtle in the ways of assimilation and intrigue, who put an end to his lesser rival Speke by accusation and never succeeded in finding the discoveries in Central Africa necessary for his own sense of glory.[16]

Burton was himself as arrogant as Baker, excusing flogging in the name of an Englishman's 'primary duty of commanding respect for himself, for his successors, and for the noble name of his nation.' He could say that African women should be seen nowhere, or in the dark, even if some of their children had the 'amusing prettiness which we find in pug-pups.' He could give up any idea of educating the African, who was like the Asiatic, 'stationary, but at a much lower level.' Yet Burton was witty and stylish and exact in his descriptions of the habits, food, drink, music, weapons, crafts, diseases, beliefs and minutiae of Africa. Nothing was too small to escape his attention or to prick his pride. The journals of his travels, which still have to serve as the written records of some primitive peoples in East and West Africa, are less an accurate picture of their nature than of the breadth of learning and narrowness of vision of the mid-Victorian mind.[17]

So the scouts of European influence were to penetrate into the interior of Africa and discover the evidence which would reassure the preconceptions of their audience. The values of civilisation were stressed, the habits of savagery were denigrated. Above all, the first explorers brought back the data for future expeditions and colonists; imperialism needs a map. As the climates and geography and peoples of Africa became a little known, so the fear of the dark continent was replaced by the complacent opportunity to exploit it. It had taken Prince Henry the Navigator of Portugal fourteen years to persuade his sea-captains to sail beyond the first great African Cape of Bojador, for these medieval Europeans had believed Arab tales that the world ended there, or that the sea boiled, or that the tropical sun could turn them all into black men. Once the Cape was passed, the slave trade could begin. In the same way, once the Victorian missionaries and explorers were to penetrate Central Africa, the soldiers and the traders could follow, if only the government would let them.

The Portuguese had begun the slave trade as a crusade, and the British were to end it as another crusade to abolish the sins of the past. Once the London Parliament declared the sale of human beings illegal in 1807, foreigners were not allowed to make money from what Liverpool had to forgo. In the first sixty years of the nineteenth century, the British always kept one or two squadrons of frigates off the west coast of Africa to extirpate the traffic in flesh. These permanent blockaders forced the coastal chiefs to give up trading in their captives and ruined the West African middlemen, except for those who dealt in the palm-oil of the Niger. The British presence in West Africa was reduced to three poor footholds of little importance.

As an investment, the continent was to be written off. Only South Africa mattered, because of its white colonists and strategic position on the route to India. The protectorate of Zanzibar counted minimally, as a convenient island that looked towards Britain's eastern empire. James Stephen of the Colonial Office wrote of Africa in 1840, 'If we could acquire the dominion of the whole of that continent it would be but a worthless possession,' while a British naval officer on coastal patrol against slavers ten years later could declare that the countries of West Africa were 'in the same state of barbarism as when they were first discovered.'[18] Even among the reformers, the ending of the slave trade led to a loss of interest in Africa. As the Governor of Sierra Leone warned in the very year of 1807: 'To abolish the slave trade is not to abolish the violent passions which now find vent in that particular direction. Were it to cease, the misery of Africa would arise from other causes; but it does not follow that Africa would be less miserable: she might even be less miserable, and yet be savage and uncivilised.'[19]

So Africa was to lapse in the European mind into a state of apathy and ignorance. The long marches and quarrels of Burton and Speke, Baker and Stanley provoked controversies, but they did not persuade the government to develop its interests in an uncommercial interior. Livingstone's far wanderings and lonely death made church groups dream of a black Christian Central Africa that could deny the advance of Islam from the north; but Whitehall would not permit the army to follow the Bible or the explorer. There were occasional punitive expeditions against such forest tribes as the Ashanti and such madmen as the Emperor Theodore of Ethiopia, but there was no occupation. Influence through Free Trade and gunboats was enough. Colonies were too costly. As Palmerston declared, civilisation should be extended by commerce alone in Africa. Not until the opposite routes to India were to be threatened by a Boer revolt in the south and an Egyptian upheaval on the Suez Canal in 1882 were the British to intervene in Africa with permanence and force. And then the other European powers would join in the scramble that was in sixteen years to draw nine-tenths of Africa within European colonial frontiers on maps, which would hardly affect the interior tribes of a continent so barely known.

Like penned slaves, territories would be seized by the quarrelling white powers. If Britain got Nigeria and the Gold Coast, then France got Dahomey and Germany the Cameroons. West, East and North Africa were shared out piecemeal. Britain and France were even on the edge of war over the claim to the Upper Nile at Fashoda. While the three naval superpowers took the lion's share, the remnants fell to Portugal, which was first in Africa, to Spain, to Italy and even to Belgium, which would manage to secure the great prize of the Congo through King Leopold's avarice and British collusion.

Yet the vast territories scrabbled together by the European powers in the last two decades of the nineteenth century were mere colours on the map. They were hardly occupied and largely misunderstood. If French troops and administrators did penetrate the caravan routes of the Sahara, it was because the desert was an inland sea with camels for caravels. The forest belts near the coasts were penetrable only by rivers, and down them European traders floated their tenuous influence and innate greed. Economic imperialism did not cause the partitions of Africa, for traders followed the flag, settlers went where colonies were declared, pioneers on the spot waited for the decisions made by international conferences. Until the discovery of gold in South Africa, the whole continent seemed an uncommercial and barren strategy, peripheral to the good of Europe.

One late Victorian novel would capture the irrelevance of the scramble for Africa in the wilderness which it claimed to own. Before Leopold, King of the Belgians, forced a profit from the Congo by the ruthless exploitation of the rubber trade, the ramshackle Belgian administration had to exist on a diminishing supply of ivory. Joseph Conrad, the seer of British sea-power, was haunted by the stories of horror that spread from Europe's exploits on that notorious river. In his most profound novel, *Heart of Darkness*, he began his explanation of the savagery at the heart of all men by stating that even London was 'one of the dark places of the earth.' As his narrator Marlow watched the brooding gloom of the monstrous city by day change to its lurid glare by night, he remarked that darkness had lain on the Thames 'yesterday'. What had the commander of a Roman trireme thought of the twisting river – 'sand-banks, marshes, forests, savages ... cold, fog, tempests, disease, exile, and death.' A young Roman in a toga, marching through the woods to some inland post, must have felt that 'the savagery, the utter savagery, had closed round him – all that mysterious life of the wilderness that stirs in the forest, in the jungles, in the hearts of wild men.' From detesting the unknown, the Roman would have begun to surrender to 'the fascination of the abomination.' Even his disgust would have made him powerless to resist an escape.

Of course, Marlow did not suggest that the European colonisers of Africa were like the Romans, who were merely robbers on the grand scale; but the conquest of the earth, which mostly meant taking it from those with a different complexion or a flatter nose, was not a pretty thing. It was only redeemed by an unselfish belief in the idea at the back of it, some sense of mission, some idol ... So Conrad set his story with the suggestion that the heart of darkness lay in the city and in the past as well as specifically in primitive lands. His hero Marlow had always craved to explore a great river in Central Africa, and at last he was appointed by a giant European trading company to captain a steamboat out there. On the map, the river looked like a snake, and like the serpent in the Garden of Eden, it was to lead Marlow into the knowledge of evil. He sailed out to Africa and watched the coastline slipping by interminably. He found the sight like thinking about an enigma. At one point, he saw a French gunboat incomprehensibly shelling the continent. The tiny projectiles disappeared into the colossal jungle and nothing happened because nothing could happen. The act was insane, faced with such immensity. The voyage itself was like 'a weary pilgrimage amongst hints for nightmares.'

Reaching the river, Marlow found black chain-gangs condemned for crimes they could not understand, building a railway that would rust before it worked. Other black contract labourers waited to die of sickness in an unnecessary pit. He met an immaculate accountant who had resisted 'the great demoralisation of the land.' He heard of the remarkable trader Kurtz, who kept the ivory coming from the depths of the interior, and who was reputed to be a prodigy of pity and science and progress. Marlow then marched to a station two hundred miles upriver and found his steamboat sunk. He began getting hungry and ferocious himself as the months passed while he repaired the boat, surrounded by 'a taint of imbecile rapacity ... like a whiff from some corpse.' All seemed unreal at the river station. 'The silent wilderness surrounding this cleared speck on the earth struck me as something great and invincible, like evil or truth, waiting patiently for the passing away of this fantastic invasion.'

At last the steamboat was repaired and Marlow took it up the great snake of the river. It was like travelling back to the beginnings of the world, when the big trees were kings and their gloom overshadowed all with 'the stillness of an implacable force brooding over an inscrutable intention.' The crew were cannibals and lived on rotten hippo meat. The passengers were foolish, pink, irrelevant European pilgrims. The grimy beetle of the ship crawled on 'deeper and deeper into the heart of darkness.' Nothing could be understood by Marlow on this journey back to the hidden roots of civilisation and the human heart, except for a sense of remote kinship with the roaring savages from the forests, a suspicion that they were not inhuman.

Nearing Kurtz's station in the interior, a white fog descended. Behind it there was a savage clamour of grief from the bush. When the fog lifted, the natives of the interior attacked the boat as it was nearly caught in a narrow wooded channel. The pink pilgrims happily shot rifles at the spears of the savages. The boat reached the station, which was stacked with ivory. The unseen Kurtz seemed to be the favourite of the wilderness which 'had taken him, loved him, embraced him, got into his veins, consumed his flesh, and sealed his soul to its own by the inconceivable ceremonies of some devilish initiation.' To achieve such riches, how many of the powers of darkness now claimed Kurtz for their own? He must have taken a high seat among the devils of the forest. He had even written a report on how to suppress the customs of the savages, in which he claimed that Europeans should appear as supernatural beings and do good by the power of their will. Yet his report ended with the scrawled words, 'Exterminate all the brutes!' The heart of darkness had blinded Kurtz's sense of mission, the lust for ivory had made him a ruthless killer. His house was surrounded by human heads on poles. Yet this 'pure, uncomplicated savagery' was a positive relief to Marlow from the 'lightless region of subtle horrors' where Kurtz now seemed to wander.

Soon Kurtz himself appeared on a stretcher, as emaciated as an ivory carving of death. He was surrounded by his tribe of African killers. They stayed on the bank, leaving him to come on board the steamboat, now piled with his elephant tusks. But he staggered back to a last unspeakable native ceremony, only to be stopped by Marlow who tried to break the spell over him, 'the heavy, mute spell of the wilderness – that seemed to draw him to its pitiless breast by the awakening of forgotten and brutal instincts, by the memory of gratified and monstrous passions. This alone ... had driven him out to the edge of the forest, to the bush, towards the gleam of fires, the throb of drums, the drone of weird incantations; this alone had beguiled his unlawful soul beyond the bounds of permitted aspirations [to] his own exalted and incredible degradation.'

Marlow took the dying Kurtz back down the river on the steamboat, while 'both the diabolic love and the unearthly hate of the mysteries it had penetrated fought for the possession of that soul satiated with primitive emotions, avid of lying fame, of sham distinction, of all the appearances of success and power.' The banks of the river slipped monotonously past the dirty boat that was the forerunner of change, of conquest, of trade, of massacre, of blessings. Kurtz died with an expression of intense and hopeless despair on his face and the words, 'The horror! The horror!' That was his final judgement upon the adventures of his soul on earth. Marlow fell desperately ill, but he survived to recognize Kurtz's extreme vision as a belief, a truth, a victory over the grey fight with death that was the fate of a civilised and moderate man.

So Marlow returned to a meaningless urban world where he had to tell a white lie to the girl who loved Kurtz, that his last word was her name. The truth would have been 'too dark altogether.' By the time Marlow told the story on the night Thames, it had lost its horror for him and merely lingered on 'impalpable, like a dying vibration of one immense jabber, silly, atrocious, sordid, savage, or simply mean, without any kind of sense.'[20]

Conrad's *Heart of Darkness* would be the best analysis of that jumble of instinct and fear with which Victorian men confronted the savage inside themselves and on their travels. No Englishman could deny the fact that Boadicea had impaled the population of Roman London on stakes or burned her captives in giant wicker men. The Victorians were not removed from barbarians by many centuries. The veneer of civilisation was as thin as the rusty plates on the bottom of the steamboat crewed by cannibals pushing up a muddy river to an unknown fate. The futility of technology, when confronted by a dark continent, was always apparent – the gunboat shelling the jungle, the rusting railway, the pilgrims uselessly gunning down natives for sport. The truth was the horror, which Kurtz saw at the base of all human desire and mission – the outcast's urge to be both god and beast, to rend, hoard, kill, dominate, indulge, pursue, satisfy and die at the limits of power and lust. The escape into total savagery and the black forests of Africa and the human self was the final solution for the prodigies of Europe such as Kurtz. Their weaker followers were left to tell the little lies of civilisation and to forget the horror which is at the root of all instinctive action in men. The European occupation of parts of Africa would ever be threatened by fear of the black, of the unknown, of the savage. And in the rain forests, fear could reach the heart of darkness.

One painter of genius was to recognise that labyrinth of guilt with which some Europeans viewed the slave trade after it had been condemned. In 1840, J.M.W. Turner painted *The Slave Ship* or *Slaves Throwing Overboard the Dead and Dying – Typhoon Coming On*. It represented a true incident, when sixty years before a slaver had thrown overboard his cargo of sick Africans to claim the insurance. The last of the captives had jumped of their own choice into the liberty of the furious seas. In the painting, the strong waves and gory sunset predominate, although sharks feed on the chained limbs of a black corpse. The slaves and the ship are almost swallowed up in the outrage of nature. The cruelty of humanity has offended the heavens, which rage at man's injustice to man. John Ruskin once owned the picture, declaring that Turner's immortality would be founded on it; but he sold it to an American buyer because it was too distressing to live in front of such a spectacle of human suffering.

Such recognition of evil and repugnance at past brutality, in which the sea predators represented the white slaves feeding on their bound prey, and in which the divine anger threatened all in a typhoon of wrath, was nevertheless disguised as a storm in the ocean. Victorian sensibility would accept so much, just as in Géricault's *Raft of the Medusa*, the black figure overhanging the other survivors is withdrawn in his distress, while the rescue boat sails indifferently by. A code governed the horrors that could be depicted. Conscience was left to interpolate what was only suggested to the eye.[21]

So the savage was enslaved, and the enslavement was reluctantly explained, although it could not be excused. A fascinating comparison between North American and South African history has defined the frontier as more of a phase of race relations than a wilderness where settlers were free to build new societies.[22] In that sense, 'savage' is a more useful term than 'frontier', with which to examine the complexities of the encounters between primitive and more civilised societies, and between human beings and their environment. Both 'frontier' and 'savage' apply in North America and South Africa to the European myth that the settlers were occupying free or nearly empty land. Actually, there was a large Indian population in North America which was thinned by disease to a small minority on the subcontinent while the black population in Africa always outnumbered the intruders. White South Africans, indeed, have always been afraid that the 'frontier' or 'savage' process may reverse itself, and the native peoples in a near encounter may sweep them from the land. Their dependence on local labour, which never applied to the Indians of North America, led to laws of separation from the larger population and attitudes of superiority to it, rather like the Jim Crow state laws against the black worker in the Southern United States. In North America, however, the frontier in the West was closed by the dominance of the European settlers and of industrial capitalism, although the savage was to become an increasing factor in the dealings of men with the wilderness in future times of conservation and green consciousness. In South Africa, the frontier was also closed, but it has never been reopened by the black majority. And there the savage has ever been present.

5
THE MYTH OF THE SAVAGE

IF THE INFLUENCE of Europe forever changed the Americas and Africa and Asia, their influence also began to change Europe. The urge to discover a place beyond the oceans or the mountains, where there was peace and joy and heaven on earth, made the new discoveries as much dreams as exploitations. The wish to find a terrestrial paradise was an inspiration behind the pens of explorers and the theories of philosophers and poets. To civilised communities, the hope of a golden age existing among simple people was almost as great as the fear of the onslaught of the savage and the beast. And once again, this desire reached back at least as far as Gilgamesh.

When the hero of Uruk physically sought for everlasting life beyond the ocean of the dead, he discovered Utnapishtim, the Noah of the Sumerians, who had survived the Flood and had begotten the human race from his sons. Utnapishtim not only reconciled Gilgamesh to the fact of death, but he also lived at his ease, lying on his back at the mouth of the great rivers. There at Dilmun, 'the lion did not devour, the wolf did not rend the lamb, the dove did not mourn, there was no widow, no sickness, no old age, no lamentation.' It was a paradise at the world's end like the later Elysian Fields of the Homeric Greeks, 'where living is made easiest for mankind, where no snow falls, no stormy winds blow, and there is never any rain.'[1] Lucian later added inexhaustible wine and honey and music and scent to the Elysian Fields, making them fit for hedonists rather than for heroes.

The unknown could be hopeful as well as fearful. Plato deliberately created his mythical Atlantis as a mirror of the vices of Greek society. The vision of the perfect state of his *Republic* was translated in the *Timaeus* and the *Critias* to the sunken island of Atlantis, with its convenient canals and excellent water system. Atlantis has been more useful as a promoter of dream worlds than as a description of a society. Yet it continued the Mesopotamian tradition of Dilmun, the heaven on earth at the mouth of the rivers, and it was to bequeath to Sir Thomas More the tradition of writing of perfect communities that existed nowhere. The Arabs, of course, had their belief in a Garden of Allah, but it had no terrestrial existence except in the secret rites of the Assassins, whose leader Hassan-al-Sabah was meant to have created an earthly Garden of *houris* and drugged bliss to seduce his followers into their political missions of death.

Another tradition lay behind the dream of a tangible paradise – the traveller's tale. Where movement was dangerous and strangers rare, the fantastic had great opportunities for embroidering the facts. The Middle Ages of Europe hardly distinguished between legend and geography, between hyperbole and ethnography, between myth and fact. Thirty years after the discovery of the New World, Cortes could be instructed to look in Mexico for strange humans with flat ears or dog-like faces. Both Eden and El Dorado were considered to be discoverable in the New World, and the name of the ocean that separated it from Europe came to be called after Plato's dream of Atlantis.

The inventor of Utopianism, Sir Thomas More, was a man who may have preferred a good joke and a good cause to his own life, but he combined Platonic tradition and the traveller's tale to create his masterpiece. His *Utopia* of 1516 is derived from a Greek word, meaning Noplace, and it is a pun on another Greek word, meaning Wellplace. It is set circumstantially and geographically by an ancient mariner called Hythlodaeus or the Nonsense-talker, who claims to be one of the twenty-four sailors left behind by Amerigo Vespucci at the fort at Cape Frio in 1504. This factual incident is the beginning of his voyage to find Utopia, an island described more precisely than Atlantis – and equally absent from the maps.[2]

Clearly More believed enough in the myth of the classical golden age to set Utopia in the New World. He claimed, indeed, that twelve hundred years before his time, some Romans and Egyptians were shipwrecked in his ideal land, and that the Utopians immediately adopted their best ideas, as they adopted the best ideas of the first Europeans whom they encountered. They were noble men and women, capable of old virtues, unspoilt by new vices. Although More makes his explorer confess that there was much to condemn in the actual New World and that many mistakes had already been made there, yet he also asserts that its primitive society and its pagan rules could help in reforming Europe. The Americas, with or without a Utopia, represented the lost innocence of urban civilisations. If their ignorance was not always bliss, it was wise because it was receptive. Folly was to be a proud and unbending European.

Not only were 'the savages' of America the product of the imagination and the policy of Europe in the New World, but also accounts of these savages, prejudiced as they were by the myths of Europe, were used to support the theories which Europeans wished to prove about the nature of men before civilisation. What Potosí was to the Spanish treasury, the American savage was to the European philosopher, an inexhaustible mine of glittering speculations. The discovery of America coincided with the Renaissance in European thought. The humanists, who found a Golden Age in the classical past, sought to find that Golden Age also in the present, in the games of children, in the songs of the people, in unspoilt rural life, and in the habits of primitive tribes before the coming of civilisation. Thus, to a certain extent, the New World was the ready-made Utopia of the Renaissance. The luck of discovery gave a geography to wish-fulfilment. The actual descriptions of Amerigo Vespucci were the inspiration for the book of *Utopia*.

One reader of More, the Spaniard Vasco de Quiroga, actually founded two Utopian communities for the Indians in Mexico, one near Mexico City and the other near Michoacan. The ordinances for these were drawn up in 1535, the year More was beheaded in England. Thus Quiroga began the translation of European myth into social fact in the New World. He was convinced that the Indians had been in the same golden state of innocence as the people in the traditional Age of Saturn, before the brutality and corruption of the age of iron had encountered them. His two hospital-villages were moderately successful, creating groups of monkish Indians who were still at the beginning of the seventeenth century 'living together in communities and devoting themselves to prayer and the pursuit of a more perfect life.'[3]

Thus the theme of the Noble Savage was not an invention of the eighteenth century, but a rediscovery of classical nostalgia. The life of the savages of the New World seemed to be the proof of the Golden Age of the Latin poets. Montaigne's *Essays* particularly demonstrate a hankering after Utopian innocence in the American woods. In his famous comparison of cannibal Indian tribes with civilised Europeans, he praises the purity and simplicity of societies living in the forests without money or private property or institutions. The ideas of a Utopia and of the Noble Savage are closely allied. Reason, unsullied by avarice or ambition, seemed at its best in a state of nature, free from urban vice. The very naked beauty of the Indians reminded Europeans of the statues of Greece and Rome and of the virtues of Sparta.

Although Renaissance Utopias moved away from Platonic models, they did idealise the savage at peace in his primitive anarchy. If Campanella and Francis Bacon and Andreae in their seventeenth-century versions of Utopia, *Civitas Solis* and *Nova Atlantis* and *Christianopolis*, praised a more technological and scientific society, they took pains to extol the virtues of the American 'savages', from the state socialism of Peru to the communal tribalism of the Iroquois and the Algonquins and the Hurons, beloved of the French Jesuits. Grotius would even argue that these hardy and warlike and generous North Americans were descended from Tacitus's German tribes, so similar were their virtues.[4] Yet just as to Renaissance humanists the classical tradition predetermined the nobility of the American savages, so to the first Spanish settlers the necessities of colonial control predetermined the bestial nature of these same Indians. What was light in Europe was darkness on the spot.

The most subtle vision of the interaction between the Old World and the New in the seventeenth century is not to be found in the works of contemporary philosophers or travellers or historians of New Spain. It lies in *The Tempest*. In the same way that Amerigo Vespucci's voyages inspired Sir Thomas More, so the wreck of the Virginian expedition of 1609 on the island of the Bermudas inspired William Shakespeare's last play. Of the two accounts of the survivors, one was called *A Discovery of the Bermudas, Otherwise Called the Isle of Devils*.[5] It stressed the magical properties of the uninhabited islands, which were held to be 'a most prodigious and enchanted place, affording nothing but gusts, storms, and foul weather; which made every navigator and mariner to avoid them as Scylla and Charybdis, or as they would shun the Devil himself.' In fact, the island abounded with fish and game, and the climate was healthy and pleasant. In all, the land was as 'merely natural, as ever man set foot upon.' The survivors rebuilt their ship and set off for Virginia. Their reports of the island led soon afterwards to its colonisation and to its spoiling.

Shakespeare also began his play with a shipwreck, then showed that the storm was raised by Prospero, a magician cast away on an enchanted island with his daughter. Prospero was once the Duke of Milan, but his dukedom had been usurped by his brother with the help of the Spanish conquistadors at Naples. In turn, this refugee from Europe had usurped the power of the native ruler of the island, whose name was Caliban. Shakespeare was fond of conceits and puns, and Caliban's name seems to derive from the first bestial savage met in the new world – cannibal Carib man. Prospero has reduced Caliban to slavery, and the dialogue between them shows Shakespeare's ambiguous view of colonisation. Caliban is foul and misshapen in Prospero's eyes – a freckled whelp, hag-born. The magician treats him as a slave and labourer, to be tormented into service. Caliban complains that Prospero has stolen his inheritance. When the magician first came, he was kind to Caliban, until he had been shown all the secrets and foods of the island. Then he penned Caliban into a rocky reservation, taking control over all else. Prospero replies that he had to dominate because Caliban tried to violate his daughter and people the island. The native is an

> Abhorred slave,
> Which any print of goodness wilt not take,
> Being capable of all ill! I pitied thee,
> Took pains to make thee speak, taught thee each hour
> One thing or other ...[6]

So Shakespeare presented the coming of colonialism. First, the policy recommended in Hakluyt's *Instructions* to the second Virginian colony, a gentle approach to the natives, until they reveal the resources of the new land. Then the condemnation of the natives as bestial and the enslaving of them as labourers. Then the confining of them on the worst land, claiming that their foul nature makes them capable of the rape of the colonists' daughters. Then the teaching of a new language, which was the instrument of domination. What had the words of Christianity done, after all, but order the Indians to obey the priests and to work their own land for the Church and the state? And so Caliban's only revenge is to use the speech he has learned to curse and to plot rebellion and to wish the white man's disease, the smallpox, on the immigrant magician who has disfigured him and displaced him, enslaved him and enthralled him in his lost paradise.

Shakespeare went on to explore the myth of the New World. Although Prospero has caused a storm to wreck his enemies on the island and to avenge himself for the wrongs of Europe, the old Counsellor Gonzalo tells the castaway King of Naples what he would do, if he were settling a plantation.

> I' the commonwealth would by contraries
> Execute all things; for no kind of traffic
> Would I admit; no name of magistrate;
> Letters should not be known; riches, poverty,
> And use of service, none; contract, succession,
> Bourn, bound of land, tilth, vineyard, none;

> No use of metal, corn, or wine, or oil;
> No occupation, all men idle, all;
> And women too, but innocent and pure;
> No sovereignty . . .

Here other courtiers interrupt to say that Gonzalo, after all, does want to be a king and rulers often forget at the end of their rule the good intentions of their beginnings. But Gonzalo persists with his vision of Utopia.

> All things in common nature should produce
> Without sweat or endeavour: treason, felony,
> Sword, pike, knife, gun, or need of any engine,
> Would I not have; but nature should bring forth
> Of its own kind, all foison, all abundance,
> To feed my innocent people . . .
> I would with such perfection govern, sir;
> To excel the golden age.

This is not Prospero's nor Shakespeare's view of the New World; they are concerned with human treacheries and frailties and the original sin of a Caliban. The shipwrecked Europeans try to assassinate each other for the gain of titles and estates, while their servants continue the torment of Caliban. As the jester Trinculo points out, the English would not give a farthing to help a lame beggar, but they would give ten times that sum to see a dead Indian. The drunken butler Stephano calls Caliban a savage and a monster, but introduces him to the delights of wine, which makes Caliban think him a god. This 'celestial liquor' is a shrewd method of debauching and controlling the native, who licks Stephano's foot to get his grog. Caliban sees freedom in exchanging masters. He rebels from the harsh discipline of the learned Prospero, and he plots with the pair of shipwrecked servants to kill the magician. If Prospero's books are burned, the three can live in sottish and animal bliss on the island. The renegade Europeans and the natives can, like William Bradford's evil Morton in Pilgrim Massachusetts, take over as the twin Lords of Misrule with their wild servant.

The rebellion confirms to Prospero that Caliban is 'a born devil', incorrigible and unteachable. He hunts down the three plotters with the most feared beasts of the Indians, 'divers Spirits, in shape of dogs and hounds.' Their names are Mountain, Silver, Fury and Tyrant – all references to the profitable Spanish despotism in Mexico and Peru. The rebellion is chased away, the plotting Europeans all forgiven, Prospero's daughter so ravished by the sight of fresh white faces that she says,

> How beauteous mankind is! O brave new world,
> That has such people in 't!

So *The Tempest* ended with the coming of the Europeans blessed by the eye of the innocent, treachery absolved by the magic of reason and the mercy of Providence, and the 'thing of darkness' Caliban abandoned to the rule of his inheritance – and even he swears to 'be wise hereafter, and seek for grace.' Shakespeare had examined many of the myths and stories of the early expeditions to the Americas – the Golden Age, the treatment of Indians, the contradictions between the natural state of man and his government, the sins and virtues that the European brought overseas in their beliefs and in their baggage.

A nice comment on the moralities of *The Tempest* soon came from the Bermudas, when a minister wrote back to England in 1615 urging more settlers to come and join the budding colony. The bad reports of the islands were merely part of God's plan to keep such an earthly paradise for the English; but the settlers must 'above all things have a care to leave their sins behind them and come hither as it were into a new world to lead a new life.'[7] They came, however, with their old sins into a new place, which did not regenerate them. Shakespeare had opened his play with one lot of the castaways planning to murder the others; he had closed it with the Europeans, temporarily repentant, leaving the New World, which only in their absence might regain its natural state through the grace of heaven. Salvation there could come direct from God without the whips or the words of civilisation. Even Prospero at the end of *The Tempest* gives up his magic powers and begs the English audience to free him from his base island and his failures there, and to return him to Europe.

> And my ending is despair,
> Unless I be reliev'd by prayer,
> Which pierces so, that it assaults
> Mercy itself, and frees all faults.
> As you from crimes would pardon'd be,
> Let your indulgence set me free.

In previous plays, Shakespeare had made the dark savage an object of admiration and pity, most particularly in *Othello*, where the murderous Moor of Venice is the credulous victim of the plotting of his

European subordinates. And even in *Titus Andronicus*, when Shakespeare did create a black villain in Aaron, the lover of the Queen and Empress, he is given a ferocious verbal attack on her two white sons, who are trying to kill their bastard half-brother:

> He dies upon my scimitar's sharp point
> That touches this my first-born son and heir ...
> What, what, ye sanguine, shallow-hearted boys!
> Ye white limed walls? Ye ale-house painted signs!
> Coal-black is better than another hue;
> In that it scorns to bear another hue;
> For all the water in the ocean
> Can never turn the swan's black legs to white,
> Although she lave them hourly in the flood.
> Tell the empress from me, I am of age
> To keep mine own; excuse it how she can.

Outside such Shakespearean subtlety, the influence of the idea of the savage on European philosophy in the seventeenth century was simplistic. The praise of men living in a state of nature in Hooker's *Ecclesiastical Polity* was perverted to justify the Elizabethan State Church. Locke largely ignored the travel literature in his personal library and preferred to quote Hooker as an authority on the fact that men did once live in a state of nature and were bound by the laws of nature. His *Essay on Civil Government*, which was to serve as the philosophical base for the Constitution of the United States of America, was imbued with the Utopian idea that a primitive man was obliged by the dictates of natural law and by the persuasion of reason not 'to harm another in his life, health, liberty, or possession.' Locke set aside the known facts of the bloody Indian wars of tribe against tribe, where he argued for his innocent version of the aborigine. Although his treatise was full of dozens of references to the American Indian, they were merely uses of the known existence of the Indian in the backwoods, in order to locate European theories of property and the validity of contract in actual place and time in the evolution of civilised society. Above all, Locke was concerned with proving that labour gave property its value and that the Indian, by not improving the soil, had a worthless stake in his preserves. For all of Locke's concern with the virtues of the state of nature, he was a propagandist for that lust after property which was the basis of the European's self-justification for dispossessing the Indian.

Since Locke was the employee of the Whig rebels under Charles the Second and the justifier of the Glorious Revolution of 1688, he was also bound to find a right of rebellion in the mythical contract that set up civilised society and in the mythical existence of natural laws. Thus he had to refute the contradictory use of the American savage made by the most ruthless and logical of all the English philosophers, Thomas Hobbes. Hobbes had wished to prove the need for total authoritarian rule under his Leviathan. Therefore, he needed to prove that life in the state of nature was chaos. America was a convenient site for him to locate his state of anarchy and continual war, which necessitated the coming of Leviathan. Hobbes first defined the state of war in primitive society in one of the greater sentences of the English language; he then sited that state of war in America, making his philosophy precede his prejudiced version of anthropology.

> In such condition, there is no place for industry; because the fruit thereof is uncertain: and consequently no culture of the earth; no navigation, nor use of the commodities that may be imported by sea; no commodious building; no instruments of moving, and removing, such things as require much force; no knowledge of the face of the earth; no account of time; no arts; no letters; no society; and which is worst of all, continual fear, and danger of violent death; and the life of man, solitary, poor, nasty, brutish and short ... It may peradventure be thought, there was never such a time, nor condition of war as this; and I believe it was never generally so, over all the world: but there are many places, where they live so now. For the savage people in many places of America, except the government of small families, the concord whereof dependeth on natural lust, have no government at all; and live at this day in that brutish manner ...[8]

So the two great philosophers of seventeenth-century England cited the American Indian to prove their philosophical case. Locke's version of the virtuous Indian was to be more useful to the Age of Enlightenment, which needed a basic belief in the goodness of human nature and of men in a state of nature to justify its assault on feudalism and the Ancien Régime. Curiously enough, it was the fashionable study of the state of nature and the American Indian which encouraged Montesquieu – that other important philosopher for the American Constitution – to read his Tacitus and to consider early Germany, when European tribes also lived in a state of nature in the woods just like the noble redskins.

The virtuous Germans, whom the Roman historian hymned in the forests as an example to the corrupt Roman citizens, were resurrected by Montesquieu as an example to the corrupt urban Europeans. Montesquieu found the origins of the free governments of northern Europe in the geography of northern Europe, in its mountains and forests which were reproduced in the northern colonies of America. He used the American Indian to support many of his theories, that the abundance of nature allowed primitive men to remain hunters and isolated savages, while barrenness made them become pastoral and barbarian nations; that the virtues of the Indians pointed up the vices of Europe; and that the sufferings of the Indians at the hands of the Spanish proved what barbarians the Spaniards were. So America, because it was a New World containing primitive people, served as the point of contrast for European vices and virtues. No comparison between nature in the raw and civilisation in clothes could be more opposed than that between the Indian and the courtier at Fontainebleau and Versailles.

So the philosophers of Europe continued to pervert the realities of the Americas for their own purposes. The Comte de Buffon found it necessary to state that all species of animal in America *without any exception* were smaller and weaker than similar strains in Europe. Even imported European livestock degenerated. The wilderness was so hostile to living things that the savages of the New World were becoming impotent. Buffon explained the large and ferocious wild beasts of the Americas by stating that they were all forced into size and ferocity by the extreme cold. The bear swelled due to the pressure of ice, the python due to its cold blood, only man shrank and could not expand his numbers to fill the empty continent, so that few savages awaited the coming of the European immigrants. The very youth of the New World showed that living things there were younger and weaker than European strains. But the American thinker Thomas Jefferson would have none of such indecision. To refute Buffon and show the vigour of the New World, he felt the need to assert 'that the proofs of genius given by the Indians of North America place them on a level with Whites in the same uncultivated state.'[9]

Whatever the European mind wanted to believe of primitive man, there was always an opinion of the Indians to bear out his belief. Only Rousseau, perhaps, of the major philosophers of the eighteenth century, was neutral about the Indian, in order to show up the exaggerations of Hobbes and Locke. In his *Discourses on Inequality*, he accused Hobbes of importing the vices, and Locke the virtues, of European civilisation into the American wilderness to prove their cases. Rousseau's man in a state of nature was neither good nor bad, merely amoral and, above all, isolated. The state of nature was a state of solitude, of wants immediately felt and satisfied, of simple needs simply met. It was neither a state of war nor of love, merely a state of existence. Yet Rousseau's version of the American Indian, though admirable in its neutrality, showed no more comprehension of the realities of tribal life than did those of Hobbes or Locke. He fell into the exact pitfall in which he saw his rivals trapped, the description of the state of nature without experiencing that state. His version of the Noble Savage may have become notorious as the European myth of the wild man of the woods; but it was as far from the truth as Geneva was from Dakota.

Descriptions of the Indians were originally recorded by the more sophisticated of the traders who first encountered them. These traders were sharply aware of the different versions of the tribes current in Europe. As Henry Hastings Sibley wrote after trading in Minnesota, the truth lay somewhere between the Indian as 'the impersonation of the chivalry of olden times, proud, hospitable and gallant' and the Indian as 'revengeful, implacable and bloody minded.' Edwin Denig supported him, recognising 'two sets of writers both equally wrong, one setting forth the Indians as a noble, generous, and chivalrous race far above the standards of Europeans, the others representing them below the level of brute creation.' Both traders might pride themselves on a true knowledge of the Indian; but they, too, sometimes fell into one or another of the extreme versions of the Indians for fear of 'the savage' and the wilderness. Henry Boller, another trader, best confessed to this rejection and acceptance of the exaggerated notion of the Indian:

I could 'paint' you ... two pictures:

The One would represent the bright side of Indian Life, with its feathers, lances, gayly dressed and mounted 'banneries', fights, buffalo hunting, etc.

The other, the dark side, showing the filth, vermin, poverty, nakedness, suffering, starvation, superstition, etc. *Both would be equally true – neither exaggerated, or distorted; both totally dissimilar!*[10]

These balanced accounts of meeting the Indians to do business in furs and to live among them were not published until modern times, when an interest in accurate social history and ecology had grown. Frederick Jackson Turner's version of the frontier was challenged: 'What the Mediterranean Sea was to the Greeks, breaking the bond of custom, offering new experiences, calling out new institutions and activities, that and more, the ever retreating frontier has been to the United States ...' Western history was now seen as another stage in the expansion of European influence across the globe. The American frontier did not change the intruder. The savages and animals and environment were overwhelmed as they always were by the invasion of industrial and urban civilisation, which brought its culture and its domination in its account books. 'The trains were the spearheads of the frontier.' As in *Gilgamesh*, the city conquered the wilderness.

Eventually the Indians were given a new voice. Their testimony was recorded by Europeans and witnessed a religious harmony between them and nature, which had been disrupted by the intrusion of European commerce, weapons, animals and parasites. The Indians of the forests and the plains had only killed enough beasts for their needs and had felt integrated with the rest of creation through their beliefs and clans. But the advent of the fur trade and the meat market, the rifle and the horse, alcohol and disease, had made the Indians hunt beaver and fox and buffalo to the edge of extinction along with the white trappers. Faced with the demands of trade for an overkill, they lost their worship and respect for their quarry. From contact with European money and parasites – for the Indians believed that animals spread smallpox and typhoid rather than human interlopers – the 'savages' ended their truce with nature and joined in the war against animals.

Before the European had crossed the Atlantic, the Indians had possessed what is now called ecological wisdom. The land was not empty, but preserved by them. 'If anyone asked Indians what they thought about animals, trees, and mountains, they answered by talking about the powerful spiritual beings that were those things. No European, whether he be Christian, rationalist, Jew, or deist, could possibly believe in ideas like those [which] enabled the Indians to live in and to change the American environment without seriously degrading it.'[11] As Luther Standing Bear declared, when he understood the concepts and speech of the European invaders, 'Only to the white man was nature a "wilderness" and only to him was the land "infested" with "wild" animals and "savage" people. To us it was tame. Earth was bountiful and we were surrounded with the blessings of the Great Mystery.'[12]

Given a voice and an understanding as Caliban was by Shakespeare, the American Indian learned to mourn his dispossession and the loss of his pact with nature. The extraordinary testimony of the Oglala Sioux mystic, *Black Elk Speaks*, expressed a lost relationship with the cosmos as well as a lament at its ending. 'You see me now a pitiful old man who has done nothing,' Black Elk of the animal name concluded, 'for the nation's hoop is broken and scattered. There is no centre any longer, and the sacred tree is dead.'[13] But only in the twentieth century was the Indian able to riposte as Caliban did to Prospero. In the seventeenth century, the only knowledge that the mass of readers had of encounters with the savages of the American forests had come from tales of capture by the Indians – always popular in early colonial society. They were the first distinctive form of American literature in English. They began with Captain John Smith's *True Relation* of 1608, and *The General History of Virginia, New England and the Summer Isles*, which he published sixteen years later. Although his account of his capture by the Indian chief Powhatan and of his rescue by the chief's daughter Pocahontas may have later strained the credulity of Henry Adams, it set the precedent for a long-lived literary genre.

Captain Smith was a far more sophisticated observer of the natives than William Bradford, and he noted that the Virginian Indians wore the skins of wild beasts, but were not wild beasts themselves They had human characteristics, being inconstant, crafty, timorous, ingenious, cautious, covetous, malicious and 'all savage'. Yet no place in Virginia was so savage that the Indians did not possess a religion, deer hunting, and bows and arrows. All things capable of hurting the Indians induced their adoration, from lightning and fire to muskets and horses. But their chief god was the Devil, whom they served more from fear than love. 'They say they have conference with him, and fashion themselves as near to his shape as they can imagine.'[14] The Indians seemed to Captain Smith to accept evil as much as the Puritans repressed it.

The prisoners of the Indians returned with their horror stories of life in the forests. All their accounts were red with massacre and torture, but they were differentiated in their moral purposes. Christian endurance and divine vengeance were stressed in the early versions of war between the northern colonists and the Indians. Edward Johnson called the first published history of Massachusetts *Wonder-Working Providence* and thanked the Lord for his extermination of several hundred Indian men, women and children at Mystic, Connecticut. The Indians retaliated in King Philip's War, raiding two-thirds of the towns in Massachusetts and committing the worst of atrocities. In the end, King Philip himself was quartered and hung on four trees, proving 'the Providence of God'.[15] After the Deerfield Massacre, one of the few survivors, the minister John Williams, praised himself for not kissing the Cross when ordered to by a Jesuit-trained Indian, who declared that Williams was 'no good Minister, no love God, as bad as the Devil.' His survival and return seemed to him like a redemption from hell.[16]

Later accounts of capture by the Indians lost the moral indignation and stressed the sensational aspects of the experience. The most interesting of these was Mrs Mary Jemison's account of her abduction by the Shawnees in 1755. She wrote with detachment of the slaughter of her family and the killing of an English officer by winding his intestines round a tree, while praising her first Seneca husband as a man of nobility and gentleness, and his tribe as full of fidelity and honesty and chastity.[17] In these tales from the wilderness, only the extremes were shown, as defined as the bars of sunlight that occasionally broke through the leaves to contradict the darkness of the forest floor.

Such stories inspired the novelists of Europe, who saw in the Indian the heroic virtues and in the American woods the shadowed corridors of passion. Chateaubriand himself visited North America and found simplicity and dignity in the Indians, but his romance *Atala*, about a Christianised Seminole maiden and her lover Chactas, owed more to his dreams of 'the new Eden' of French Louisiana than to the facts of the local tribes. Only in Gustave Doré's later illustrations of that immensely successful novel were the labyrinths of the backwoods shown in their full terrors and complexities to the European mind.

In the Leatherstocking novels, James Fenimore Cooper turned the tales of Indian abduction into literature. If he did not capture the ambiguities of the life of the man in the woods, he did understand the confrontations and the conflicts. His Leatherstocking is based on the career of Daniel Boone, the legendary frontiersman always withdrawing into the forests further and further ahead of the axes of civilisation, finding 'a population of ten to the square mile, inconvenient.' As the American settlers advanced, the aged Leatherstocking finally joined the displaced Indians on *The Prairie*, the terminal treeless wilderness of the time.

Leatherstocking remained the anarchic figure of the woods with his archetypal defiance of Judge Temple in *The Pioneers* over the new game laws, setting independence against authority, personal morality against overall legality, freedom against houses. As the frontiersman said of himself, he had spent 'five years at a time without seeing the light of a clearing, bigger than a wind-row in the trees', yet he would back his own sense of right and wrong before any church deacon's. The judge sentenced him to jail, then paid his fine – the law followed by mercy. The conflict was not resolved, although the future was clear. As the judge's daughter exclaimed: 'The enterprise of Judge Temple is taming the very forests! How rapidly is civilisation treading on the footsteps of nature!'[18]

Leatherstocking, however, stayed at the frontier of the combat between the farm and the forest, the town and the trees. If he was more complex than the noble Indian of *Atala*, Cooper's hero remained too straightforward for much analysis at depth. Not until Nathaniel

Hawthorne wrote *The Scarlet Letter* were the ambiguities of the Puritan conscience properly examined in its fear of the power of the surrounding wild. In the novel, Hester Prynne was convicted of adultery and an animal surrender to lust. She was imprisoned and shunned by the Puritan township of Boston and went to live in a cottage at the edge of the 'dark inscrutable forest ... where the wildness of her nature might assimilate itself with a people whose customs and life were alien from the law that had condemned her.' Out of the forest came her old husband, once an alchemist and a captive of the Indians; he now embodied the power of guilt and the past, and he walked the mazes of revenge. Hester's lover, the Reverend Dimmesdale, was torn between his social position and his sense of sin and hypocrisy. Only in the woods, where he met Hester in secret, could he give way to his passion and seek the strength to be an outlaw like her. 'She had wandered, without rule or guidance, in a moral wilderness; as vast, as intricate and shadowy, as the untamed forest ... Her intellect and heart had their home, as it were, in desert places, where she roamed as freely as the wild Indian in his woods.' She could not persuade Dimmesdale to flee with her to the wilderness, only back to the lax cities of Europe. But during his final sermon in Boston, he died from the poison of remorse or the sorcery of her aged husband. And Hester ended in her lonely cottage as the revered adviser of the people of Boston, seeking in her courage the comfort for their own 'wounded, wasted, wronged, misplaced, or erring and sinful passion.'[19] Puritanism was no answer for the savagery of desire or for the conflicts between society and individual liberty.

Outside literature, these conflicts were condemned. Liberty was thought to exist only in the relationships of European society. Where settlers like the French in Canada had intermarried with the Indians, they were accused by such early Massachusetts novelists as Timothy Flint of losing their own identity and becoming savages. Flint was one of the first apostles of the doctrine of manifest destiny, and although he recognised the revolting details of displacing the Indian tribes, he considered the process inevitable. He did not deny 'that the white borderers have too often been more savage, than the Indians themselves.' But the basic incompatibilities between red men and white men made it impossible for the Indians to live on the same ground. 'Beside a repulsion of nature, an incompatibility of character and pursuit, they constantly saw in every settler a new element to effect their expulsion from their native soil. Our industry, fixed residences, modes, laws, institutions, schools, religion, rendered a union with them as incompatible as with animals of another nature.'[20]

The recognition of aggression within all men, of the violence of passion within every individual, did not lead to an understanding of the Indian tribes. Extermination was always a simpler policy than conversion, integration or even separation. The Spaniards, indeed, did convert their decimated Indians to the Catholic faith, while outside a few rare communities such as the praying Indians of Martha's Vineyard, the colonial Americans were always readier to use a musket on the heathen than a Bible. Even though their own wild frontiersmen and the diplomacy of England and France often provoked the Indians into assaults and atrocities, the governments of the American States would persist in regarding Indian cruelties as innate rather than as a violent response to eviction. The idea of the savage expressed in propaganda or fiction, true story or philosophical abstraction, served to prejudge primitive men to destruction by the advance of the European settler, government and city.

One masterpiece of the eighteenth century broke this pattern of conquest by the civilised. That was Daniel Defoe's *Robinson Crusoe*. He had heard the story of a castaway, Alexander Selkirk, who had been shipwrecked alone on an island for four years. On the basis of this true story, Defoe set the case of a man forced to live without society. Robinson Crusoe's virtues are his limitations. His attention to daily detail is the reason for his survival. His rite of living is the discipline of his soul. Instead of moaning about his fate, he draws a list of things. He uses the mechanical skills of Europe to make a little fortress for himself in the wilderness. He never questions whether it is worth living at all. God he cannot fight nor question, so he prepares to defend himself against wild beasts and savages. Even the good and evil of his condition are reduced to a ledger of comforts he enjoys and miseries he suffers. Morality is a mere matter of ease or pain to Robinson Crusoe alone, for he has no company.

The first thing that the castaway does is to make a pale round his tent and his cave. He drives in piles or stakes, which he later increases to a double wall and an impenetrable wood round his dwelling. After a world of time and difficulty, which passes the first year of decades of solitude on the island, he feels secure enough physically to question the purpose of his existence. As he hunts for game, his journal records 'the Anguish of my Soul at my Condition, would break out upon me on a sudden, and my very Heart would die within me, to think of the Woods, the Mountains, the Desarts I was in; and how I was a Prisoner lock'd up with the Eternal Bars and Bolts of the Ocean, in an uninhabited Wilderness, without Redemption.' His careful husbandry and daily employment are the continuing redemption of his soul, tempted by idleness. Then he realises that he may not be alone on his island at the mouth of the Orinoco river. There may be visiting savages 'worse than the Lions and Tigers of Africa', where he was once a slave to the Moors.

There are men worse than beasts. Cannibals, modelled on early accounts of the Carib Indians, come to the shores of the island to cook and eat their war captives. The finding of a footprint first scares Crusoe, then later he views the remains and the sights of the cannibal feasts. Yet Crusoe's disgust is tempered by a curious diffidence over his right to judge savages who eat one another. He is no Spaniard, whose butchery of the Indians has disgusted the conscience of the world. 'How do I know,' Crusoe asks himself, 'what God Himself judges in this particular Case?' Cannibalism is not against the conscience of the savages. 'They think it no more a crime to kill a Captive taken in War, than we do to kill an Ox; nor to eat human Flesh, than we do to eat Mutton.'

Thus Crusoe is confused about his right to attack people who do not attack him. One day, however, the bestiality of men eating one another leads him to save one of their prisoners, a native he calls Friday, and then a Christian captive from Portugal, the country which has committed so many atrocities against the Indians and the Africans. At the sight of the killing of innocent captives, Crusoe's blood finally boils, and he reckons God's wrath speaks through his musket to justify his killing of the Caribs. Yet it has taken him twenty-four years of solitary life and the reaching of old age to make him decide to assault the cannibals and give up his solitude for the companionship of another man. He justifies this yielding to the need for a comrade by stating that he has to have another man to work the boat that can take him from the island. A hermit's life is not enough in the end. Society is his siren call.

Once Crusoe has begun to communicate with Friday, Defoe makes him state the theory of the equality of mankind. Crusoe finds in Friday the same powers and reasons and affections that are found in any man. All that Friday needs is instruction, which Crusoe now gives to him, as Gilgamesh did to Enkidu and Prospero to Caliban. With another being to speak to him, Crusoe spends his best year on the island. Conversation is more of a necessity than escape. If deliverance comes, it is involuntary. A ship of mutineers puts into the island, giving Defoe an opportunity to compare the brutal behaviour of the English pirates with the Caribs – both will kill innocent captives in cold blood.

The novel of *Robinson Crusoe* is a detailed journal of survival in the wilderness, a defence of reason and method against passion and speculation, an allegory of the nature of the individual and society, and a *Pilgrim's Progress* in the guise of a castaway's record. Crusoe knows that the primary savage is himself, although he will hardly admit to it. Only work will keep his anguish and self-despair under control. His ribs contain the misery of his soul just as his pale contains his cave and his enclosure turns his wild goats into a semblance of a flock of sheep. 'I must keep the tame from the wild,' Crusoe writes about his goats, 'or else they would always run wild when they grew up, and the only Way for this was to have some enclosed Piece of Ground, well fenc'd either with Hedge or Pale, to keep them in so effectually, that those within might not break out, or those without break in.'[21] He has built a Pale for himself too, so that he might not break out nor others break in upon him.

In the end, however, Crusoe cannot reconcile himself to the total loss of society. He rejects the solitude of the wild state. As he rescues Friday, then the Portuguese captive, then the captain of the mutinous ship, so he makes them take an oath of loyalty to himself as master and governor of the island. His solitary liberty ends in tyranny over his little society. His first scruples against intervention become the philosophy of the conquistador. Finally, with the admirable irony of Defoe, the primitive Friday is taken by Crusoe to Spain, where he distinguishes himself by killing wolves and bears in the civilised country of imperialism. For Friday is close to the innocent beast himself, and he has the measure of the animal. And at the very end of the book, Crusoe sends supplies from Brazil to the quarrelling colonists of his old island. His spoiled solitude has become a turbid plantation. No more than *The Tempest* is a wedding fantasy is *Robinson Crusoe* a children's escape story. Both of these works are also commentaries on the encounter of European reason with savage custom in the New World, while Defoe's delicacy of conscience would have made Las Casas delight in his defence of the Indians from unprovoked attack.

Africa's influence on Europe before the eighteenth century was minimal. Chiefly, the profits from the slave-trade built the docks and port cities of England and Holland and France. But Africans themselves in Europe were few and far between, and little was known of the continent itself. As Swift wrote:

> ... Geographers in Afric-Maps
> With Savage-Pictures fill their Gaps

And o'er inhabitable Downs
Place Elephants for want of Towns.

What the African experience did encourage was geographical determinism and a feeling of superiority. On the title pages and in the illustrations of the new works of descriptive geopraphy, the continents were associated with ideas: 'Europe – crowned, cuirassed, holding a sceptre and orb, with weapons, scientific instruments, a palette, books and Christian symbols; Asia – garlanded and richly dressed, holding an incense-burner, and supported by camels and monkeys; Africa – naked, with elephants and lions, snakes and palms, and often with the sun's rays like a halo on the head; America – naked, with a feathered head-dress, holding a bow and arrow.' Just as the iconography of the ancient Egyptians had despised the Nubians, so Christian art despised African. The age of the expansion of Europe excited a crude complacency. As Samuel Purchas boasted in his *Pilgrims*, Europe had embraced the inferior glove and had mastered the universe. 'The quality of Europe exceeds her quantity, in this the least, in that the best of the world ... Asia yearly sends us her spices, silks and gems; Africa her gold and ivory; America receiveth severer customers and tax-masters, almost everywhere admitting European colonies.'[22]

The easy victories won by the handfuls of Europeans in their wooden boats against large and powerful African kingdoms seemed inexplicable unless there was an innate superiority in the European. Christianity, of course, appeared the primary advantage, with its convenient text that condemned the black sons of Ham to serve the white sons of Japhet. If also man was made in God's image, and there was only one God, then that image must be the white man's image, of which the black man was a lowly copy. When priests like Father Cavazzi met African pride in the kingdom of the Congo, Matamba and Angola in 1687, they found it incredible. 'With nauseating presumption these nations think themselves the foremost men in the world, and nothing will persuade them to the contrary ... They imagine that Africa is not only the greatest part of the world, but also the happiest and most agreeable.' And this belief came from a people 'more animal-like than reasonable.'[23] Brute instinct could be the only cause of refusing to recognise European dominance.

The very exuberance of tropical Africa was held to be a reason for the backwardness of local society. De Golbéry echoed centuries of false European assumptions about black communities when he asserted of the West African that he was both slothful and sober, agile and indolent, existing 'in the sweetest apathy, unconscious of want, or pain of privation, tormented neither with the cases of ambition, nor with the devouring ardour of desire'. His needs were few and easily supplied, while 'those endless wants which torment Europeans' were unknown to him.[24] De Golbéry thought that the harshness of the European climate was the reason for its economic success. Barrenness bred invention and trade, just as fertility bred sloth and happiness. Clinical observations of the deterioration in the performance of Europeans in West Africa and the West Indies seemed to bear out the idea that climate led to human degeneration. 'There is, in the inhabitants of hot climates,' as one doctor in the tropics observed, 'unless present sickness has an absolute control over the body, a promptitude and bias to pleasure, and an alienation from serious thought and deep reflection.'[25] Cold apparently provoked philosophy, heat led to hedonism.

The European philosophers of the Enlightenment, indeed, were no friends to the African. Linnaeus's major classification of plants in his *Systema Natura* of 1735 also included a racial classification based on skin colour, roughly placing a white race in Europe, a red race in America, a yellow race in Asia and a black race in Africa. From these divisions and categories, presumptions about the superiority of the European race could easily grow. While both Voltaire and Rousseau could suggest that black people were mentally inferior to white people, David Hume could go further, stating that there had never been 'a civilised nation of any other complexion than white, nor even any individual eminent either in action or speculation. No ingenious manufacturers amongst them, no arts, no sciences ... Such a uniform and constant difference could not happen, in so many countries and ages, if nature had not made an original distinction betwixt these breeds of men.'[26] By the time Linnaeus had split mankind into two species and Lord Monboddo had included orang-outangs in the second species and Edward Long had claimed in his *History of Jamaica* that Europeans and Negroes belonged to different species, the efforts of European thinkers to justify the mistreatment of Africans as animals and savages had reached a peak of logical folly. The cult of reason did not help. For good behaviour was held to derive from teaching, not from feeling nor instinct. 'Pity is not natural to man,' Dr Johnson declared. 'Children are always cruel. Savages are always cruel. Pity is acquired and improved by the cultivation of reason.'[27]

So the African was condemned by his lack of education to a position near to the beast, although he did benefit in a minor way from the cult of the Noble Savage. Mrs Aphra Behn's extraordinary novel about an

African prince called Oroonoko was a popular and theatrical success in the seventeenth century and gilded the chain of the slave-trade. Mrs Behn was naturally careful to make her black hero a prince, and thus separate him from the common run of Africans. Although the opening of the novel in Africa was full of outrageous rhetoric, Aphra Behn had lived in Surinam and she knew of the conditions of slavery. Oroonoko was lured aboard a slave-ship, sold and reunited with his lost love Imoinda, still unexpectedly as chaste as ever. He led a slave revolt and had to kill Imoinda to keep her from white lust. He himself was nursed back to health, only to be tortured to death as a dire warning to other slaves.

Oroonoko was presented throughout as a refined and magnanimous man, a courageous hero with a charisma of his own. Even as a slave, 'people could not help treating him after a different manner, with designing it. As soon as they approached him, they venerated and esteemed him.'[28] If Aphra Behn suggested that his qualities were princely, she seemed to make no distinction between white and black. In fact, his primitive virtues were more than those of the Noble Savage, they were those of the Roman hero. He played the same role for the seventeenth century that Chateaubriand's Chactas played for the eighteenth century, the classical lover of the Golden Age. As the slave trade increased, European romantic writers found it more convenient to set their nostalgic tales in the uncontaminated American forest rather than in the looted African jungle.

Even more timely for Christian Europe was setting the virtues of humility in potential slaves. Europe's religious faith continued to praise meekness without practising it. There was a genuine respect for the simplicity and obedience of an Adam and Eve in the Garden of Eden, forever lost to city-dwellers, but perhaps retained by bush Africans, who might well be both the perfect flock for missionaries and the servants of the new plantations. If Prester John, the mighty Christian monarch of Africa, was a myth, then the black character might prove to be the untilled field from which the word of God might spring up a hundred-fold, in simple submission. If the diversity of the West Coast tribes in nature and culture was known to the local slave-traders, as was the resistance of many of the tribes to Christianity, the conscience of Europeans was comforted by a belief in the 'childlike' African, being guided through centuries of servitude into the way of the Cross. Rare was the evidence of such men as Charles Wheeler, who spent ten years as an agent of the Royal Africa Company and declared: 'We Christians have as many idle ridiculous notions and customs as the natives of Guinea have, if not more.'[29]

So Africa influenced Europe hardly at all before the age of industrialisation made factories greedy for tropical commodities. Africans were rarely seen in Europe and easily assimilated. It is a curious fact that the fifty thousand Africans in England in the eighteenth century disappeared into the white population, forming no ghetto of their own. Racial prejudice was already explicit in the 'scientific' classification of species; but popular prejudice in a Protestant country was far greater against the Pope than against a black. While polite London society welcomed the redeemed African 'prince', Job ben Solomon, the London masses cheered the black American boxer Molyneux in his two savage encounters with the English champion, Tom Cribb. Outside the plantations of the New World, Africa and Africans were looked upon with some sympathy, unknown people from an unknown continent, ready to be civilised, hardly understood.

When Marco Polo had claimed that no other man, Christian or pagan, Tartar or Indian, had known and explored so many parts of the world since the time of Adam, his claim might well have been correct, even if his account of his travels had largely been disbelieved. Medieval Europe simply had not credited that a civilisation as advanced as China truly existed. Polo's accounts of Kubilai Khan had seemed more fantasy than fact, making a Munchausen out of a Mongol prince. What Europe had known of the terrible attack of the Golden Horde was in total conflict with Polo's description of the high civilisation of Cathay. Thus China had remained an Eastern dream during the early period of Ming isolation, until brief contact was established through Macao and Canton under the Portuguese and the Spanish. Although Mendoza wrote an account of Chinese history based on traders' tales, not until the journals of the Jesuit Matteo Ricci were published in 1610 was any accurate description of Chinese civilisation available in Europe. Such inventions as printing and the manufacture of paper had not been ascribed to the Chinese, merely the manufacture of lacquer and tea and other oddities.

Ricci had lived in Peking as an astronomer and mathematician on the fringes of the court, and he had begun a Jesuit policy of trying to convert China to Christianity through astronomy. As the Chinese emperor was considered to be the mediator between heaven and earth, and as the imperial calendar set the times of harvests and irrigation and festivals, the right ordering of the lunar calendar was all-important. Although Chinese astronomy was quite exact, the new instruments of Europe developed for sea navigation were superior. Ricci could use them to confound his Chinese rivals and to impress the emperor and the mandarins. As he wrote, the Chinese would accept foreign inventions, if they found them better than their own. Their only problem with Europeans lay in 'an ignorance of the existence of higher things and from the fact that they found themselves far superior to the barbarous nations which surround them.'[30]

Ricci's successor, Adam Schell, however, became the chief court astronomer to the last of the Ming emperors and transferred his skills to the first of the Manchus. Although he was disgraced by his enemies on a charge of treason by encouraging the growth of Catholicism in China, 'a tiger that will lead to future disaster', his successor Ferdinand Verbiest again proved that Western astronomy was more accurate than Chinese calculations. Thus Jesuit influence persisted at the highest levels of Chinese society, as did the Jesuit hope of converting the whole of China to Catholicism by imperial edict. It was a vain hope, for no Son of Heaven would accept the authority of the Pope at Rome. Yet the illusion did not die with Verbiest. Jesuits acted for the Chinese in drawing up the border treaty with the Russian barbarians in 1689, and they were not driven out of China until the Society of Jesus was itself abolished in the 1770s and late Manchu China decided to persecute Christianity, despite the technical skills of its missionaries.

Unlike America and Africa, which could only provide versions of the Noble Savage to mirror the vices of Europe, China could provide an example of a counter-civilisation. Both Voltaire and Adam Smith used China to castigate their own societies, while the decorative techniques called *chinoiserie* were applied to furniture and china and even pagodas in English gardens. But an understanding of Europe in China was biased and suspicious. In fact, by the opening of the nineteenth century, Western influence in China was again tied to Portuguese Macao and to the foreign traders' strip at Canton. There the foreign merchants were unable to gain recognition even by local Chinese officials; they were forbidden to use Chinese in their communications which had to be in the form of 'humble petitions'; they were treated officially as 'barbarinas'; and they were confined to the malarial and overcrowded suburb of the city. Lord Amherst's mission to Peking in 1816 would end in total failure, with the Chinese emperor rejecting the clocks and other mechanical presents of the West as valueless foreign objects, and instructing the distant King George the Third to 'progress towards civilised transformation.' At the opening of the age of industrialisation, a Great Wall of self-sufficiency surrounded all of China, and its influence on Europe was minimal. Contact between the two peoples was unnecessary and futile. Even during the Ming Empire's brief period of expansion by sea, a Chinese captain had advised with contempt that all barbarian beings could be treated 'like harmless seagulls' and that barbarian peoples need not be feared more than 'touching the left horn of a snail.' In his opinion, the only real dangers for Chinese sea-power were the greed of foreign merchants and the savage waves themselves.[31]

Montaigne had been the first commentator on European expansion to see the effect of its terrible mechanical civilisation on the cultures of the rest of the world. Although a sixteenth-century man, he was sceptical of the paramount virtues of Christianity and appreciative of the achievements of the great urban civilisations of Mexico and Peru. He had feared that Europe would hasten the decline and ruin of the New World by its contagion, and that it would sell its opinions and arts very dear. The cities of Cuzco and Mexico had seemed to Montaigne to be 'of awesome magnificence', and the very virtues of the Indians were the reason for their conquest by the Europeans. The Indians had, indeed, the qualities of the ancient Greeks and Romans and were only overcome by the deceits, the illusions and the technological skills of the Spaniards, 'equipped with a hard and shiny skin and a sharp and glittering weapon ... (and) the lightning and thunder of cannon and arquebuses.' The result of the victory of Europe was terrible. Was empire and trade worth the price? 'So many cities razed, so many nations exterminated, so many millions of people put to the sword, and the richest and most beautiful part of the world turned upside down, for the traffic in pearls and pepper! Base and mechanical victories!'[32]

The expansion of Europe was merely a series of mechanical victories, in which a superior organisation of war and commerce overcame the acquiescence or resistance of the cultures of the rest of the globe. Furthermore, the mechanistic view of the world and of the universe allied with the spirit of scientific enquiry was becoming dominant among western philosophers. Descartes rejected the idea that he was a rational animal. He was a thinking being, and because animals did not think, they were not of the same order as mankind. They did not act; nature acted upon them and moved them as a clock moved the hands which told the time. Beasts were sorts of machines, because they could not talk or be intelligent. Brutes not only have a smaller degree of reason than men, but are wholly lacking in it.[33]

Although motivated by a falling apple rather as Eve had been in the Garden of Eden, Newton discovered the theory of gravity and made a hypothesis of a mechanistic universe obeying divine laws rather like an almighty engine. The cosmos was a vast celestial contrivance. Scientific method was the oil on the cogs and ratchets of the universe. The world was similarly a machine which moved by universal laws. Human reason alone was capable of discovering its secrets. The proper study of mankind was not man, but the analysis of nature. Vivisection became the rage in the seventeenth century. The followers of Descartes held that animals felt no more pain than trees did, when they were pruned. So William Harvey and his disciples killed thousands of beasts very slowly to prove the theory of the circulation of the blood, while Robert Boyle methodically suffocated kittens and frogs and snakes in compression jars to show what a vacuum could do to living things. This illustration of the divorce of humanity from nature was captured in one of the first luminous paintings of the industrial revolution, *An Experiment on a Bird in the Air Pump* by Joseph Wright of Derby. In it, children are entranced by the sight of the dying small feathered being trapped in the airless tube. Reason and investigation overrule even the wonder and pity of the young.

The Age of Reason, however, did provoke a cult of the landscape, as long as it was managed properly. The great parks of England were then created. Carefully wrought, they had to seem uncontrived and were designed on the principle that 'nature abhors a straight line.'[34] The poet Alexander Pope, who planted his own Picturesque garden outside his riverside villa at Twickenham, condemned a formal scene, in which:

> No pleasing intricacies intervene,
> No conflicts to perplex the scene ...
> The suffering eye inverted Nature sees,
> Trees cut to statues, statues thick as trees ...

He set out his point of view in *Of Taste*:

> Consult the genius of the place in all
> That tells the waters or to rise or fall ...
> Now breaks, or now directs, the intending lines,
> Paints as you plant, and as you work, designs.[35]

This was a vision of the contrivances of man as embracing the harmony of nature. The savage would be trimmed from sight without spoiling the qualities of creation. The body of the earth would be massaged, but not raped. The beast would be tamed and appreciated, but not mutilated and massacred. The curate of St Mary's at a village in southern England, Gilbert White, proved to be the founder of field ecology when he wrote his observations about his parish in *The Natural History of Selborne*. He discerned the mutual dependence of all chains of life on one another. Nature is a great economist, he declared, as it converted 'the recreation of one animal to the support of another!'[36] Although Providence ruled the natural order, a helping hand from man could improve things. Yet in White's vision, creation was one whole structure and ruled by God. Man's intervention was limited and should only be within the modes of God's works.

The Natural History of Selborne did not accept the prevailing laws and economics of the age, in which enclosures and harsh criminal laws and the industrial revolution were repudiating the communion between country people and the natural world. Goldsmith's celebrated poem, 'The Deserted Village', was more true of the time than White's pastoral Arcadia in Hampshire. Faced with landlords grabbing the commons and factory machines killing home crafts, the villages were being abandoned, while the moral virtues 'leave the land' for America.[37] A ferocious penal code 'cherished the death sentence' for poachers of petty animals: to kill a hare or a deer was a hanging matter.

In a century of habitual cruelty to animals, human possessors were as cruel to their fellows whom they dispossessed. Even the enlightened Earl of Shaftesbury, the arbiter of the manners of the time, thought that the vulgar of mankind stood in need of 'such a rectifying object as *the Gallows* before their Eyes.'[38] Public executions attracted vast crowds, although the baiting of bulls and bears in pits and gardens was less popular. William Hogarth's moral cartoon, *The Four Stages of Cruelty*, demonstrated how common was brutality to beasts throughout the population. Dogs and cats and birds were tormented, horses and sheep and donkeys were abused. The moral was, however, a product of the Age of Reason, which Dr Johnson would have approved. Cruelty to animals led to cruelty to man and punishment by the law. The savage Tom Nero of the engravings began by thrusting an arrow up a dog's behind and proceeded to cut the throat of his pregnant mistress and ended after his hanging on a marble slab, dissected like a pig's carcass by the knives of surgeons. Human beings were more bestial to their own kind than to the rest of creation.

* * *

THE MYTH OF THE SAVAGE

As the Industrial Revolution destroyed the village economy to replace it with factory employment in the growing cities, the separation between the human and the natural world increased. The engine age, indeed, introduced the mechanics of colonial exploitation at home. There was little difference between the plantation owner with his overseer and slave labour in neighbouring huts, and the factory owner with his foreman and wage labourers in surrounding slums. The outwork system of hand-loom weaving had itself been a proto-colonial system; but when it was replaced by the centralised factory with its hundreds of spinning jennies and mules, a colonial revolt was provoked. The Luddite disturbances in Georgian England were, in their way, a home-grown version of the resistance of the Aztecs to the Spaniards, turbulent and doomed protests against the base and mechanical victories of the society to come.[39]

The Industrial Revolution did, however, change the attitudes of Europe towards savagery. Two important differences developed. Firstly, the urban mob which had terrified visitors to London in the Gordon Riots and had carried through the French Revolution in Paris became widespread in such new factory cities as Birmingham and Manchester, Lyons and Milan. The growth of the cities themselves set up a barrier between the new suburbs and the new slums as profound and actual as the abyss of misunderstanding between the village and the forest, the literate and the primitive. Secondly, a Romantic revolution against the dark Satanic mills of the industrial way of life shook apart European literature and set it on an ambiguous – and even decadent – admiration of the wild, the subversive and the savage. The revolt of the Luddites of the provinces may have been a real one against the high price of bread and the shortage of jobs, yet their national influence was minimal. It was the inchoate and Romantic outbursts of the anarchic group of writers associated with the philosopher Godwin and the Shelleys and especially Lord Byron which broadcast an international rebellion in manners and sensibility that was to subvert the Age of Reason and glorify a permanent attack on the values of a mechanical and commercial society.[40]

Although Byron himself was to become the hero of the intellectual Luddites and of the cult of the savage in European thought, Godwin and his daughter Mary Shelley were the chief propagandists of the external conflict between reason and revolt. Godwin's own novel, *Caleb Williams*, was influential on the whole rebel expatriate group of Byron and the Shelleys and their friends. It was considered both disturbing and seditious. It explored a split personality, who was represented by two men, the rational Williams and the demoniac Falkland. Williams felt bound to prove that Falkland was a murderer, while Falkland both persecuted him and shunned him, fleeing away among the rocks and precipices, and showing an 'inconceivably, savagely terrible anger.' Yet both men despised and attacked brutal or bestial men. They were bound together by society in a master and servant, tyrant and slave relationship, in which both innocence and guilt were confounded. Godwin seemed to say that only obsession and the power of will could break through the laws of men into an unholy freedom.

Godwin's influence lingered round the Lake of Geneva, where Mary Shelley conceived her *Frankenstein* in 1816 under the influence of her husband Percy Bysshe Shelley and Lord Byron. While a novel of horror, *Frankenstein* was also a penetrating analysis of the revolt from reason, of the conflict between the Satanic urge for the outer limits of freedom and the need for the restraints of society. The arguments of Rousseau and Hobbes resounded. The hero Count Frankenstein was himself a compound between Shelley and Byron, and he sought the origins of life and human nature. He created a monstrous image of himself that was a foul caricature of his desires. 'His limbs were in proportion, and I had selected his features as beautiful. Beautiful! – Great God! His yellow skin scarcely covered the work of muscles and arteries beneath; his hair was of a lustrous black, and flowing; his teeth of a pearly whiteness; but these luxuriances only formed a more horrid contrast with his watery eyes; that seemed almost of the same colour as the dun white sockets in which they were set, his shrivelled complexion and straight black lips.'

In fact, Frankenstein had created the savage and distorted version of himself, the fiend of the inward dream. This monster raged away, and yet he was innocent. Created by a man, he looked for the company of men, who could not bear his sight. His distorted features predetermined their fears, just as the painted bodies of primitive tribes struck fear into their enemies. But Frankenstein's monster did not choose to be deformed. He was created abhorrent to men. When he later confronted Frankenstein on the wild summit of a Swiss mountain, he was full of bitter anguish, disdain and malignity. His unearthly ugliness was almost too horrible for human eyes, and yet he rightly complained to his master: 'Remember, that I am thy creature; I ought to be thy Adam; but I am rather the fallen angel, whom thou drivest from joy for no misdeed. Everywhere I see bliss, from which I alone am irrevocably excluded. I was benevolent and good; misery made me a fiend. Make me happy, and I shall again be virtuous.'

So Frankenstein's monster presented the paradox of the Romantic revolt and the noble savage. If man was born good, why did he behave so badly? In the case of the savage, his uncouth appearance caused his rejection by civilised men. This rejection warped the behaviour of the savage into malice and revenge. The monster had heard of the discovery of the American hemisphere and had wept 'over the hapless fate of its original inhabitants.' He could not comprehend that man could be at once so virtuous and magnificent, yet so vicious and base. But in his own case he was no Adam, but an outcast demon. 'God, in pity, made man beautiful and alluring, after his own image; but my form is a filthy type of yours, more horrid even from the very resemblance. Satan had his companions, fellow-devils, to admire and encourage him; but I am solitary and abhorred.'

So the monster put the case of those who were doubly deformed, the first time by creation, the second time by society. What of the black slave, whose colour made him seen evil to his white masters, and whose work on an American plantation denied him family life? What of the slave of the new mines or factories, made filthy and broken by his toil with wages too low even to keep himself? What of the convict, condemned by his character as well as his crime to wear out his life in a cell or the Australian desert? The monster forced Frankenstein to make him an equally monstrous mate, or else he would smash the engines of science that had made him so deformed. If he got a mate, they would go to the vast wilds of South America where the beasts would be their only companions.

Frankenstein began to make a female monster in the desolate Orkneys. Then he considered that her creation could breed a race of devils which might make 'the very existence of the species of man a condition precarious and full of terror'. If science created the damned of the earth, they would revolt and destroy the society that damned them. At the last moment, Frankenstein cut up the female monster, and in revenge, the monster destroyed Frankenstein's own wife and his best friend. Now Frankenstein felt compelled to hunt his monster to the uttermost ends of the earth and destroy him 'as a beast of prey.' His need for revenge devoured his soul. He died in vain in the wastes of the north, while his monster mourned him and left to kill himself on a funeral pyre. The savagery in man had killed the science in man, but now turned upon itself. When the Luddites and the damned had destroyed the technocrats, what might they do but destroy themselves?

The later hold of the Frankenstein legend on the mass mind through the dream machine of the cinema paid tribute to its original prophecy of the conflict between the demands of the machine age and the atavistic urges of passion. The monster was the noble, warped, revengeful savage in the heart of Count Frankenstein himself, whose Promethean urge to discover the secrets of human life merely loosed the wild in his soul. He believed the origins of existence lay in light and Adam, but the flaws in his own nature released a demon of the dark. He learnt too late that the conflict between human and engine, mercy and massacre, good and evil, science and sabotage, must be fought to the final destruction or to life without end, amen.

If the novel of *Frankenstein* unconsciously mirrored the conflicts of the civilisation to come, Byron himself spread wide the cult of the savage, the Corsair or Mazeppa, the fierce pirate or the Cossack outside the restraints of usual morality. By his person and pen, Byron made the outlaw respectable and notorious. Nothing more strangely represented his admiration for the primitive man, whom he could never be, than his curious eulogy of Daniel Boone in his world-weary and final poem, *Don Juan*. There the cult of the savage reached its apogee through the English poet, whose lordly way of life contradicted his values, and whose wasted death made him the symbol of a freedom that he never lived nor possessed. Byron was always a peer in life, if a bandit in spirit. His Daniel Boone was conceived in an Italian villa at an extreme remove from the American forest and the author's style of life. Personal luxury dreamed of a savage Utopia.

> Of the great names which in our faces stare,
>   The General Boone, back-woodsman of Kentucky,
> Was happiest amongst mortals anywhere;
>   For killing nothing but a bear or buck, he
> Enjoy'd the lonely, vigorous, harmless days
>   Of his old age in wilds of deepest maze ...
> 'Tis true he shrank from men even of his nation
>   When they built up unto his darling trees,
> He moved some hundred miles off, for a station
>   Where there were fewer houses and more ease ...

Yet Byron's Boone was not all alone, but the progenitor of a new race. This 'sylvan tribe of children of the chase' lived in a young unawakened world where there was no sorrow.

> The free-born forest found and kept them free
> As fresh as is a torrent or a tree ...
> Simple they were, not savage; and their rifles,
> Though very true, were not yet used for trifles ...
> Serene, not sullen, were the solitudes
> Of this unsighing people of the woods.[41]

Byron continued to state, however – so much for Nature. Civilisation had both its variety and its joys, such as 'the sweet consequence of large society, war, pestilence, the despot's desolation ... ' The best answer to the Malthusian and mechanical tragedy to come was his own Romantic revolt against the age of overpopulation, even if it meant an exaggeration of the threatened pleasures of life in the woods. In Byron's own phrase about Boone's children, 'simple they were, not savage', he already prophesied the return of the word 'savage' to its meaning of wooded rather than bestial, of desirable rather than horrible, of blessed rather than cursed.

Byron also in 1823 praised the earthly paradise found in the South Seas by Bougainville and Captain Cook. His curious poem, *The Island* or *Christian and His Comrades*, was based on the mutiny of the Bounty; but it again painted the idylls of a wild life, where the Polynesians were 'the naked knights of savage chivalry.' This image of uncontaminated virtues developed the versions of earlier European commentators, who saw in the first accounts of the South Seas what Montaigne had seen in the first account of the Americas – the Germans of Tacitus risen again, barbarian societies turned by their primitive life into paragons of rural goodness. The Byronic savage, however, was gloomy, restless, bold and individualistic, seeking glory and self-expression – the inner lust of all men after their own perfection, a ceaseless striving of nature to break out of the chains of civilisation. With the coming of the factory to the city and the monstrous regimentation of urban labour, such fearful liberty became a banner of revolt, and the forest began to seem a refuge rather than a threat.

6

THE WILD VERSUS THE MACHINE

'THE FIRST INHABITANTS of the world knew not the use either of wine or animal food,' Mr Escot declared at dinner in Thomas Love Peacock's *Headlong Hall*:

> It is therefore, by no means incredible that they lived to the age of several centuries, free from war, and commerce, and arbitrary government, and every other species of desolating wickedness. But man was then a very different animal to what he now is: he had not the faculty of speech; he was not encumbered with clothes; he lived in the open air; his first step out of which, as Hamlet truly observes, is *into his grave*. His first dwellings, of course, were the hollows of trees and rocks. In process of time he began to build: thence grew villages; thence grew cities. Luxury, oppression, poverty, misery, and disease kept pace with the progress of his pretended improvements, till, from a free, strong, healthy, peaceful animal, he has become a weak, distempered, cruel, carnivorous slave.

Other guests, who believed more in the Age of Reason than in this Romantic view of a primitive being, protested that there was no comparison between the animal life of a wild man of the woods compared to the felicity of a Newton or a Locke, a Shakespeare or a Milton. But Mr Escot did not agree:

> On the score of happiness, what comparison can you make between the tranquil being of the wild man of the woods and the wretched and turbulent existence of Milton, the victim of persecution, poverty, blindness, and neglect? The records of literature demonstrate that Happiness and Intelligence are seldom sisters. Even if it were otherwise, it would prove nothing. The many are always sacrificed to the few. Where one man advances, hundreds retrograde; and the balance is always in favour of universal deterioration.[1]

THE WILD VERSUS THE MACHINE

Headlong Hall was written the year of the final defeat of Napoleon at Waterloo, when its author Peacock was constantly seeing Shelley and his friends and engaging in their discussions about the Romantic revolt of the feelings from the rule of Reason. Although the idea that mankind had deteriorated from a golden age and had become bestial by eating meat was satirised, Peacock was serious enough to continue his strictures about the contrasts of savage life in his second novel, *Melincourt*. In that, Mr Forester buys a baronetcy for a friend, Sir Oran Haut-Ton, who has been caught very young in the trees of Angola. 'He is a specimen of this natural and original man – the wild man of the woods' – an *orang-outang* – 'a genuine facsimile of the philosophical Adam,' whom the Greeks used to worship as the God Pan or a Satyr. Sir Oran becomes a Member of Parliament of the borough of Onevote, and uses his silent strength and sense of honour to save damsels in distress. Although an accomplished musician on the flute and French horn, he lacks speech, which is an artificial faculty in Mr Forester's opinion. Sir Oran obviously understands human language, but merely chooses not to reply in it. Mr Forester and his creator would hardly have been surprised to find a chimpanzee called Washoe speaking the deaf-and-dumb sign language called Ameslan a hundred and fifty years later. By then, the close relation between men and apes had been proved.

The relationship between humans and animals now became a matter for the law in Britain. There was an increased sensibility on the issue, not only among visionaries such as William Blake with his impassioned poems:

> A robin redbreast in a cage
> Puts all heaven in a rage,
> A dove-house filled with doves and pigeons
> Shudders Hell through all its regions.
> A dog starved at his master's gate
> Predicts the ruin of the state.
> A horse misused upon the road
> Calls to Heaven for human blood ...

In 1809, a former Lord Chancellor called Erskine tried to persuade the Houses of Parliament to pass a bill securing the legal rights of animals. It was ironical that, after the emancipation of the slaves a little while before, animal freedom was considered more important than female liberation; but Lord Erskine had been moved more by Thomas Taylor's *Vindication of the Rights of Brutes* than its predecessor, Mary

Wollstonecraft's *Vindication of the Rights of Women*. The preamble to his proposed law repeated the words of *Genesis*, which gave Adam the power over the rest of creation:

> Whereas it has pleased Almighty God to subdue to the dominion, use and comfort of man the strength and faculties of many useful animals and to provide others for his food; and whereas the abuse of that dominion by cruel and oppressive treatment of such animals is not only highly unjust and immoral, but most pernicious in its example ...

According to Erskine, abuse of domestic animals hardened the heart against the natural and benevolent feelings of humanity. Although his bill was lost, another one was passed thirteen years later, making it an offence cruelly and wantonly to 'beat, abuse, or ill-treat any horse, mare, gelding, mule, ass, ox, cow, heifer, steer, sheep or other cattle.' Because of the national love of hunting and pursuing and baiting wild 'game' for sports, the categories of protected beasts were limited. Two years later in 1824, the Society for the Prevention of Cruelty to Animals was formed in Old Slaughter's Coffee House in St Martin's Lane in London. Among its founders were leaders in the crusade to end human slavery.[2]

This major step forward in Britain was not yet matched in the independent United States of America. Curiously, however, a cult there was growing of the innocent Adam of the forests of the New World set against the cruel decadence of the Old. This archetype was seen as the original man before the Fall: many American writers and painters still held that another Garden of Eden still existed in the backwoods. 'Here's for the plain old Adam,' Emerson was to write, 'the simple genuine self against the whole world.'[3] This biblical and American hero was threatened by the advance of civilisation and the machine. The virgin land was being raped and tilled and fenced. 'What the world of America is coming to,' Leatherstocking again declared in *The Pioneers*, 'and where the machinations and inventions of its people are to have an end, the Lord, He only knows ... How much has the beauty of the wilderness been deformed ...The Lord has placed this barren belt of prairies behind the States, to warn men of what their folly may yet bring the land.'[4]

American artists of the nineteenth century entangled ideas of God's nature and of God in nature. Their visions of America as a primordial wilderness, the Garden of the World, and the original Garden of Eden in Paradise were confused with two other versions of the Garden, one primitive, the other pastoral. No conflict was seen between the forest and the axe, between the prairie and the railroad when it was built. In creating a garden, even a market garden, in the wilderness, the New World might find itself a Paradise Regained that the Old World had lost. For the Americans, the fall from Eden was a fortunate tumble into a heaven on earth that could be recreated by themselves. It took decades for the Hudson River school of painters to conceive that the wilderness might become nature despoiled rather than Eden regained. For instance, George Inness's painting, *The Lackawanna Valley*, was to show the locomotive intruding into the Claudian composition, already broken with tree stumps and hazy with industrial smoke.[5] De Tocqueville's earlier lament had reached the landscape artists. There was a melancholy pleasure in seeing the American solitudes and 'the savage's natural grandeur' before triumphant civilisation should displace them. 'One is in some sort of hurry to admire them.'

There were different perceptions of the role of the savage within the Garden and the wilderness. The Indian simply did not appear in the view of an empty country waiting for the pioneers, as in Bingham's painting of Daniel Boone leading settlers over the Cumberland Gap. Nor did he appear in Jewett's *Promised Land* or Bierstadt's *Oregon Trail*: the West was a vacuum for the taking. But in the vision of the wilderness as an open and free place in which to escape from the artifices of civilisation, the Indian was the embodiment of natural liberty, particularly in the extraordinary paintings by George Caitlin among the Pawnees and Comanches and Choctaws and Mandans and Sioux. The Indian represented barrier and murderer only in the creed of manifest destiny, by which the original American stood in the way of onrushing greed for gold, of ranchers for cattle.

Whether the American Adam was seen as frontiersman or Indian, he was still given dominion over the rest of creation. Only in the beginnings of the classification of animals and plants in the wilderness was a certain respect engendered for them. When Thomas Jefferson became President, his interests in natural history were among his motives in sending the Lewis and Clark expedition across the uncharted West: the Corps of Discovery returned to Washington with several hundred plant and animal specimens. Some of these went to grace the only working natural history museum in the new nation, that in Philadelphia of Charles Wilson Peale, who even painted himself raising the curtain on his long Hall with the mastodon's skeleton and the thousands of exhibits, arranged according to the Linnean system. 'Such a museum,' he declared to Jefferson, 'easy of access, must tend to make all classes of people in some degree learned in the science of nature without even the trouble of study.'⁶

Enlightenment in the science of nature was necessary for most of the peoples of the earth. Thomas Jefferson himself did not see the anomaly of studying creation and yet being a slave-owner himself. His concern over other species did not extend to emancipating his fellow creatures, although the British Parliament had already abolished the slave trade. That peculiar and terrible institution was to haunt the American landscape and conscience until the Civil War. The swamps of the Southern States, indeed, became associated with the wrongs of slavery, which was held to poison society. Edgar Allan Poe might symbolise the situation in his pestilent and noxious tarn infecting the whole House of Usher; but later Abolitionist writing positively identified a lingering disease, 'a *forest* of moral evil more formidable, a barrier denser and darker, a *Dismal Swamp* of humanity, a barbarism upon the *soil*, before which civilisation has thus far been compelled to pause.' To Harriet Beecher Stowe, the wild and dreary belt of swamp-land around the Southern States was an apt emblem in its exuberant vegetation 'of that darkly struggling, wildly vegetating swamp of human souls, cut off, like it, from the usages and improvements of cultivated life.'⁷

A Civil War would be necessary to liberate the American slaves, and they were only to be freed by the superior technology and war machines of the Northern States. For the century had been that of the industrial revolution and the spread of economic and biological imperialism. The world market and the new machinery and the sea-borne parasite were exploiting all distant cultures and other species and environments. And at the same time that the technological superiority of the industrial revolution was ready to overwhelm the world, European thinkers began to construct an ideology that would confirm in words the triumph of the machine. If the Crystal Palace Exhibition of 1851 heralded Victorian supremacy in engineering, the same decade declared a new philosophy of race and patronage that was to put the savage exactly in its place. In 1854, Nott and Gliddon proclaimed in *Types of Mankind* that progress came from a 'war of races', Bulwer Lytton spoke of history in terms of race to the Leeds Mechanics Institute, and the Comte de Gobineau began the publication of his infamous *Essays on the Inequality of the Human Races*. To de Gobineau, the two more inferior varieties of the human species, the black and the yellow, were the crude foundation, 'the cotton and wool, which the secondary families of the white race make supple by adding their silk.'[8] De Gobineau was even to praise Christianity as a suitable religion for the inferior African, just as it was fitting for the poor of Europe. It would teach them humility and hard work.

The conditions of the new factory system did breed a similar loss of heart in the urban poor of Europe and the plantation serfs of the globe, set to labour for the raw materials needed by the new machines. At an early date 'the city, once conceived as a representation of Heaven, took on many of the features of a military camp: a place of confinement, daily drill, punishment. To be chained, day after day, year after year, to a single occupation, a single workshop, even finally to a single manual operation, which was only a part of a series of such operations – that was the workers' lot.'[9] This system of labour resulted in what all drill systems are intended to accomplish – a loss of spirit, a degradation of life, a servile obedience, a toleration of hard times.

The English workers blackened down the coal mines were often as degraded as the niggers of the chain-gangs. When Jack London was to spend a few months as a down-and-out in the East End in order to report on the condition of the English working class at the turn of the century, he found in the poor Londoner that shiftless resignation and hopeless apathy which the Victorian imperialists claimed always made Africans inferior to white men.[10] Yet the identical East Ender, translated to a trading-post in Africa, instantly became a white overlord filled with the virtues of Empire. This showed one thing – that the same industrial system which exploited both white and black working classes produced the same broken spirit in them and encouraged only the divisions between them. After all, the London mob had been the most lusty and fearsome in all Europe during the eighteenth century; yet after a hundred years of the factory system and of the delusion of imperial grandeur, it was usually tame and loyal to crown and government.

In the same way that the savage man was often provoked into acting like a savage because he was so treated, the working class of Europe was eventually provoked into industrial and political action by being exploited as a working class. The books of Marx and Engels paralleled the books of the early sociologists and ethnologists in classifying sorts of people. Although no friend of capitalism, Marx declared that its great civilising influence lay in its rejection of the deification of nature. 'Nature becomes for the first time simply an object for mankind, purely a matter of utility.' For Marx, the particular power by which men distinguished themselves from beasts was that they produced their means of subsistence. They were tool-using animals, who could produce 'their actual material life'. Marx ignored the fact that beavers and bees and some termites also did as well in this respect as the first human hunters and gatherers. But he was trying to locate the distinction of mankind in its productive activity, which he was blinkered enough to deny to the rest of creation, although even in its wild state it produced much of the food and materials necessary for the continuing life of mankind.

If Marxist thought stuck to economic criteria and defined even primitive man in terms of his labour rather than his society or culture, the end result of the categorisation of men into proletariat and bourgeoisie was to encourage the struggle of one against the other. For the job of Marxists has always been to make the proletariat *conscious* of its role. This has meant making a working man see that he was a proletarian in order that he might help to change the state. By his consciousness of his social inferiority, his progress would become possible. Thus the class struggle could take the place of Darwin's struggle for survival in nature.

Although in 1859 Darwin limited his theory of natural selection in the origin and maintenance of species to his observations on nature and not on human society, he was both as responsible and as blameless for what the social Darwinians made of his theories as Christ was for the Christians. Marx even wanted to dedicate *Das Kapital* to him and commented to Engels: 'It is remarkable how Darwin recognises among beasts and plants his English society with its division of labour, competition, opening up of new markets, "inventions", and the Malthusian "struggle for existence".'[11] Darwin did not, but his followers did. His theory of the continual fight of man towards a fitter species echoed both Aristotle's vision of man striving towards perfectibility and Hobbes's darker view of nature as a war of all against all. Most importantly it allowed social philosophers and anthropologists to patronise different cultures and races as losers in the struggle for existence.

If Darwin can be excused for the excesses of Darwinism, he grew to know of the use which was being made of his work. In his final major book, *The Descent of Man*, published in 1871, he recalled the astonishment with which he first saw a party of wild and naked Fuegians on the bleak shores of Argentina during the voyage of *The Beagle*. He concluded: 'There can hardly be a doubt that we are descended from barbarians ... Man still bears in his bodily frame the indelible stamp of his lowly origin.'[12] Darwin eventually made the step of relating his biological work to human society, and by doing so, he had to recognise the inferior in mankind and the savage in man himself.

* * *

Early anthropology converted the theory of the savage into a classification, if not a fact. The man responsible for this demeaning of the word was Sir Edward Burnett Tylor. His seminal works, *Primitive Culture* and *Anthropology: An Introduction to the Study of Man and Civilisation*, were published in 1871 and 1881 respectively; they became important as they suited the Darwinian mood of radical contemporary thought. Tylor declared that mankind might be 'roughly classed into three great stages, Savage, Barbaric, Civilised.' Forgetting the Latin origin of the word, meaning 'wooded', Tylor declared: 'The lowest or savage state is that in which man subsists on wild plants and animals.' Man then reached the barbaric state of farming the land or herding beasts for food. Civilised life began with writing, which bound together 'the past and the future in an unbroken chain of intellectual and moral progress.'[13]

So Tylor equated the savage with the lowest state of man. As he was often credited with making anthropology into a social science, his classification was of importance to primitive people and to imperialists. Although Tylor believed in the slow cultural development of tribes by themselves, his readers took his classifications to justify the occupation of the lands of savages in order to rescue them from the lowest state and advance them through colonial control to a higher one. Thus the duty of civilising justified the fact of exploiting. Primitive men could jump from savagery to the city without an intervening period of barbarism, just as the new American people were sometimes to be credited with jumping from barbarism to decadence without an intervening period of civilisation.

Such classifications of the savage depended on the argument that economic considerations were paramount. The subtle kinship and totem beliefs of the Australian aborigine, for instance, were thought devilish and retarded because his food-gathering habits were thought beastly. Yet all primitive peoples did not need nor wish to move immediately to the doubtful benefits of the shanties of the new cities. To gather grubs from tree-bark like a woodpecker was probably more healthy than opening a can of bully beef. The trouble with the categories of early anthropology was that they were too related to the available material evidence, from flint knives to maize-pounders. Before the detailed fieldwork of such modern anthropologists as Malinowski, who still was to call his civil and sensible Trobriand Islanders 'savages' in order to provoke his readers into considering the paradox of this happy and primitive style of life, the equation of the nature of a man with his way of feeding and housing himself was degrading, limiting and untrue. The arbitrary assumption that economics defined the quality of a human rather than morality, religion or neighbourliness was merely a false feeling of superiority within a group of thinkers from a few nations which had learned to dominate the world through the machines and wealth of the industrial revolution.

Yet the early anthropologists did at least encourage men in the nineteenth century to look at precivilised peoples, however these were classified. They did not condemn savages as degenerates in the manner of Comte Joseph de Maistre, who had loathed Rousseau's idea of the perfectible primitive, spoiled only by civilisation. Natural man was not a savage to de Maistre, for man lived naturally in cities as well as in forests. The nature of man anywhere at any time was as God willed it to be. God had even willed that the American Indians, so beloved of Rousseau, were actually the brutish outcasts of previous societies. The New World was not so new, but peopled by the rejects of the Old. The savage band was hardly the noble forerunner of European society. It represented a degraded state of both human nature and human living. In fact, it was the near to the state of Original Sin. Savage races were the decayed remnants of lost or defeated civilisations, and thus they lived in the harsh conditions of Adam after his expulsion from the Garden of Eden or of the cursed tribe of Israel.

Biblical fundamentalists had carried on the reasoning of de Maistre. Using the Old Testament as a basis for anthropology, they had identified the black peoples of the world as the Sons of Ham.[14] By this theory, semi-civilised Hebrew herdsmen had been expelled into a wilderness which had led them into backsliding. As all men were descended from the Ten Tribes of Israel, such people as the pygmies of Central Africa must have degenerated. Their environment had affected their standards; the jungle had made them savage. A belief in the Old Testament did imply regress in primitive peoples. This was one of the reasons that the fundamentalists were to be so antagonised by Darwin's first findings. His discoveries would not only point back to the ape as the Victorian gentleman's progenitor, but to the savage as the Victorian gentleman's fellow.

The theory of the catastrophic creation of the earth's crust and creatures, which was elaborated by the great pre-Darwinian palaeontologist Cuvier, had also helped to explain the degeneracy of the savage. Cuvier, who was the Sherlock Holmes of the reconstruction of prehistoric monsters from fossil bones, held that a succession of cataclysms had destroyed whole animal systems and had piled up their remains one on top of the other in the successive layers of the earth.

Prehistory was the heaping of natural disaster upon natural disaster, not the slow changes brought by the millions of years of evolution. This concept of catastrophes suited the fundamentalists, who held that the biblical Flood was the last of these great global disasters, which had left behind the marine bones to be found in the slate quarries of Germany and the limestone caves of England. Cuvier also made it an article of faith that no form of man had existed before the last catastrophe. Again believers in the Bible seemed vindicated by science. Creation was recent, the Old Testament was true, and thus primitive people had degenerated from their higher origins in Israel.

The first Darwinian anthropologists immediately attacked the fundamentalists. While the Archbishop of Dublin had misused Darwin's observations on Tierra del Fuego in 1856 to try to prove that men left in the lowest degree of barbarism 'never did and never can raise themselves unaided into a higher condition', Tylor did show in his works on primitive culture that the gradual sophistication of primitive tools and tribal organisation through the centuries presupposed a development in most societies rather than a regression. Divine guidance was not necessary to explain the progress of tribes. Tylor's contemporary, Sir John Lubbock, who published *The Origin of Civilisation* in 1870 and was credited with losing a parliamentary seat because of his radical views on prehistory, considered primitive beliefs as a part of the evolution of religion, which always developed in any given society from atheism through fetishism and shamanism to idolatry and finally to monotheism. Such a view of religion seemed outright blasphemy to believers in the unique truths of the Bible. Although Lubbock's views were too original and schematic, they did point the way to an evolutionary theory of religion as a slow unfolding of the meaning of God through the centuries, rather than the thunderclap of a sudden revelation to His chosen people.

The pioneer American anthropologist, Lewis Henry Morgan, broke completely with the degenerative theory of primitive man. He published his major work, *Ancient Society* in 1877 with the subtitle, *Researches in the Lines of Human Progress from Savagery through Barbarism to Civilisation*. To Morgan, there was a universal pattern in the progress of society. The extent of property ownership chartered the spread of civilisation. While savages owned few personal possessions, civilised society was rigorous about property and the ownership of land. Although Morgan was not allowed by his religious adviser, the Reverend McIlvaine, to mention the word evolution in his work, he did launch a Darwinian attack on the idea that the words 'savage' and 'degenerate' were synonymous. His search for a universal organisation for all tribes and clans in the world under the Latin name of *gentes* was as oversystematic as Lubbock's views on the development of religions, but his detailed descriptions of the democracy of such tribes as the Iroquois proved the sophistication of many precivilised cultures with their ethical standards comparable to any in the urban world.

At the end of the nineteenth century, the discoveries of mammoth tusks carved by prehistoric men and of drawings of mammoths on the walls of cave-dwellers in France finally proved that human societies had existed with animals that were not on Noah's Ark. If the biblical account were untrue, then savage tribes need not have degenerated from the Hebrews. Ironically enough, the finding of these caves began to suggest the opposite case, that civilised man could degenerate through overspecialisation of function and unfitness for survival in changed conditions. The bones of large European cave-bears were identified in caves deep in the recesses of the mountains. Originally the bears had roamed the surface of the earth, but increasing periods of hibernation in caverns had led to a loss of size and an inability to defend themselves against predatory neanderthal men. The giant cave-bears had been the masters of Europe, until prehistoric men had learned to kill them ritually. Near Nuremberg, the skulls of the cave-bears were discovered in niches, arranged like trophies; near the Drachenloch, their thigh-bones were set in a pattern all pointing in the same direction. Next came the extinction of the Neanderthal ritualists who themselves seemed to suffer a regression, since their final skeletons were more degenerate and less adapted for survival than those of *homo sapiens* or Cro-Magnon man, our direct ancestors and the painters on the walls of the bear-caves.

Modern specialists on the extinction of many prehistoric species state that the last cause of degeneration is probably the excessive demands placed on the individual by new conditions of life.[15] If all the energy of a species is occupied by mere survival, its fertility declines. This is true of both animals and men. The declining numbers of the pygmies and the Australian aborigines and the Eskimos represent an inability to adapt quickly enough to European influences from the overspecialized functions of primitive hunting. Precivilised peoples, indeed, do degenerate when exposed to advanced civilisation. Moreover, sophisticated civilisations capable of gigantic engineering feats such as the Roman Empire and the tank-builders of ancient Ceylon can be overwhelmed by barbarian people. Not only does civilisation breed its own degeneracy by over-domestication like the cave-bear, but also its technical superiority may make it unfit to resist more violent societies. If degeneracy is measured by success at adapting to new conditions of life, the most comfortable of species and nations may well be held to be those said to be the most degenerate.

Strange allies of the Darwinians appeared, the writers who believed that civilisation itself was doomed. These decadents, in love with the fall of civilisation, opposed the fundamentalists in love with the Fall of Man. To the more extreme of them such as Lautréamont, there remained by 1870 'only a few drops of blood in the arteries of our consumptive epoch.' Yet he did not believe in any dream of innocence or Noble Savage, characterising the writings of Rousseau and Chateaubriand as 'odious and particular whimperings'.[16] The flight back to the primitive was not possible for a Parisian dandy. The exquisite was the enemy of the simple.

Yet the very decadence of life and art in the late Victorian age provoked a reaction from its complexities that was to make primitive simplicity seem the height of sophistication. Gauguin was the forerunner of this change of attitude. Struck by the sight of some Javanese dancing-girls at the Paris World's Fair of 1889, he first fled to Brittany to recreate a crude and barbaric style of painting. Among the fishermen there, as he wrote to his wife, he lived 'like a peasant under the name of Sauvage.' Soon he left on the first of two trips to Tahiti, where he hoped to enter into nature and live a 'free, animal and human' life. Disease and disillusion were the consequences, but also the capturing of a primitive style that was to change the perceptions of European art. As Gauguin declared in 1897 in a contradiction to the falsely simple 'classical' art of Puvis de Chavannes, 'He's a Greek, whereas I'm savage, a wolf let loose in the woods.'[17]

That Gauguin should have gone to Tahiti was proof that the last Utopia on earth now seemed to lie in Polynesia – if not for long. Although missionary accounts of the idolatry and cannibalism of the Polynesians had already blackened the image of the idyllic islanders of the South Seas, the original vision of Bougainville and Captain Cook was as certain and enduring as that of Columbus: a new earthly paradise had been found. As Joseph Banks had noted on Cook's arrival in Tahiti, 'the scene that we saw was the truest picture of an Arcadia of which we were going to be kings that the imagination can form.' The Tahitians had bodies so beautiful that they 'might even defy the imitation of the chisel of a Phidias.'[18] Even love-making with young girls was a public joy, the ground spread with flowers and the air sounding with flutes; as

Bougainville wrote, he thought that he was transplanted into the Garden of Eden or the Elysian fields, where 'Venus is a goddess of hospitality ... every tribute paid to her is a feast for the whole nation'.[19] Yet so were the parasites and venereal disease, probably brought in by the Europeans.

By the time of Cook's second voyage, he had found what Columbus had found on his second trip to Hispaniola, that the serpent was in Eden and the serpent was the European. He wrote in his journal that the British had introduced the Tahitians to wants and diseases which had destroyed their happiness. He then asked Montaigne's question, what had the natives in all of the Americas gained by the commerce they had had with Europeans?

As went Tahiti, so went the South Pacific. Whaling ships followed those of the explorers, iron axes replaced stone tools, a chief's power depended on his few imported muskets rather than his thousands of spears. The demoralisation of the South Sea Islanders was a rapid affair; a few decades turned an idyll into a debauch. The inevitable reaction was produced, the savage counterattack of the islanders on those who inflamed their greed and threatened their way of life. On Cook's third voyage, the Tahitians revealed to him that they indulged in secret human sacrifices; he himself was killed by an ambush on another island. In the same way that the Caribs had spoiled the vision of the New World by their cannibalism, so the maneaters of New Zealand and elsewhere reddened the image of the Noble Savage of the South Seas. And when the missionaries reached Polynesia in the early nineteenth century, the imposition of Christian moralism on the taboos and customs and liberties of the Polynesians blackened the appearance of the natives from carefree Adams to indolent Calibans and Lucifers. So much so that when Gauguin did reach a French-ruled Tahiti to escape bourgeois Europe and paint his barbaric dream, he found only drink and despair and a nostalgia for lost bliss, an inertia in his supine golden women, a title for a reclining nude such as *Nevermore*, an obscure death for himself in the Marquesas Islands.

The fate of the aborigines of Australia was even more swift and sad. As their friend and helper Daisy Bates was told by the Bibbulmun, they believed that the first white men were sometimes the returned spirits of their own dead relatives. Once they had discovered that the white men were only too human, they were quickly destroyed by drink and disease and contact with alien ways. Most destructive was the erosion of their taboos, which had protected them from the elements and their gods. The promiscuity of the white men, who broke the social and sexual taboos of the bloodgroup, caused a degeneration among the aboriginal tribes and left them defenceless against the white immigrants in a way that they had never been defenceless against the outback. 'The Australian native can withstand all the reverses of nature,' Daisy Bates commented, 'fiendish droughts and sweeping floods, horrors of thirst and enforced starvation – but he cannot withstand civilisation.'[20]

Early European stories of survival among the aborigines were the reverse of the captivity myth fostered by the Puritans. There was no question of the passive European victim suffering rape and torments at the hands of the Indians before his or her folk chastised the savages with divine vengeance. The male immigrant Australian actively fled towards the aborigines to escape from the prisons and miseries inflicted upon him by his white gaolers. Since Australia was founded as a convict colony, the victim was the man harshly judged by English law; his flight to the wilderness was his search for the remission of his sins through the free gifts of nature.

Although most of these European refugees into the bush could not survive the harsh conditions there, some few did. William Buckley lived for more than thirty years with the aborigines near Victoria, while the notable convict and explorer George Clarke became a chief among the Kamilaroi tribe. He entered completely into their primitive way of life, submitting to tattooing and tribal ceremony, and learning the secrets of subsistence in the wild. He was successful as a white savage, but finally he could not resist the easy pickings of civilisation and became a cattle-thief, plundering the advancing herds of the new settlers. He was captured and again escaped, recaptured and finally hanged. Along with the few other white savages who also survived their experiences, Clarke's adventures made him into more of a Morton of Massachusetts, a Lord of Misrule who had chosen the company of savages in order to indulge his own wild nature, than a Captain Smith of Virginia, seized by the natives and saved from death only by the love of a native princess.

Yet George Clarke was escaping from a tyranny far worse than any Puritan theocracy. However severe the excesses of nature might prove to those human beings who tried to live in the Australian bush, the conditions in the early penal colonies were worse. The brutality of the guards and the severity of prison life turned the convicts into brutes of toil. As one prisoner declared before his hanging after a rebellion on Norfolk Island: 'Let a man's heart be what it will, when he comes here, his Man's heart is taken from him, and there is given to him the Heart of a beast.'[21]

In his classic novel of the early Australian experience, *His Natural Life*, Marcus Clarke showed clearly that if the convicts were vile, their unspeakable life changed them into worse brutes. His concept of his hero, Rufus Dawes, as a wronged man who would not break under any torture or humiliation, made him the symbol of the continent itself, which needed the terrible dumb power of men's work and deaths to change its wasteland into a place fit for urban life. The irony of the title, which used the judge's sentence on Dawes that he was to be a convict for the term of his natural life, was exposed in the long-drawn agony of his unnatural life among the terror and filth of the penal settlements. The strength of the book lay in its grand coherence, in its comprehension of the brute labour needed to scratch out a civilisation from a land that was found to be nearly empty of mankind – and was quickly made more empty by such horrors as the Line of 1830 in Tasmania, when all the white settlers tried to form a line of beaters and to drive the last of the aborigines to one corner of the island, exterminating them like wild game on the way.

So, on the last continent to be discovered by the ships of Europe, the lines of men and fences and furrows were to overwhelm the nomadic native tribes, which had survived so barely that they could only offer token resistance to the coming of their unchosen heirs. In the building of the new Australia, however, the first brutal labour was done by a line of immigrant men, the chain gang. By importing that institution, the English colonists brought their brutality with them. As the Darwinian Herbert Spencer once wrote, savageness begets savageness. The few aboriginal tribes were unlucky that their first encounter with the white invaders was so often with the criminal and bestial among them, who translated their mistreatment at the hands of their own kind into a savagery towards aliens, still living in the Stone Age without the weapons to oppose them.

To this day, animal paintings by Australian aborigines remain a sacramental art, prayers that the departed beasts will return to their hunters in abundance – even the crocodile, which is revered for its meat and eggs. The sea-eagle is worshipped as the best hunter and taker of human souls back to the Dream-Time before the coming of the white men. This is the time which still inspires this primal art of mankind, even if God may sometimes be depicted as the ancestral Great Kangaroo.

7

THE GREAT ENGINEER

SYMBOLICALLY, THE WAR of the civilised man against the savage has always been the attack of the line against the irregular. In terms of agriculture, the straight edge of the axe and the fence hemmed in the labyrinths of the woods. In military terms, the column and the square were sent in to combat the ambushes of the forest peoples. It did not always work. Generals trained in the geometric niceties of European battlefields could have their ordered troops massacred by Indians, like Braddock, or toppled by frontiersmen, like Pakenham at New Orleans.

Yet despite the military failure of the thin red line in the American forests and deltas, geometry and reason were chosen to demonstrate the authority of man over the contours of nature. Thomas Jefferson chose to organise the American wilderness, before its details were even mapped, on a grid. The new United States of America was not even a country, only a commercial coast fronting an unknown inland ground. Yet Jefferson used his mind to pattern the obscure interior of the Louisiana Purchase into squares, 'so laid it out despite the fact that he was a countryman who understood curving fields and wandering streams and the vagaries of plantings; laid it out to our eternal complication, his grid the woe of his fellow citizens ever afterward, impossible to find those damned cornerposts, impossible to track those inhuman squares. A Dionysian land, a Dionysian continent, plains and tortuous rivers and sharp mountains and wandering shores, and he overlaid upon it as Euclidian a grid as if Apollo truly reigned and the Furies were forever banished.'[1]

So the lines conceived in the mind of man tried to impose themselves on the irregularities of nature and the savage. Wherever men were confronted with an unpredictable, eroding situation, they tried to order it into the balloon-frame house with its tidy cubes of rooms, into the oblong chest or curved hogshead for the transport of supplies, into the notched tally-stick or ruled ledger for the counting of goods. When

Locke had made his assertion in his chapter 'Of Property' that 'in the beginning all the world was America,' he had meant that the soil of the continent was worthless until it became part of a system of economics. To cultivate crops and to stock cattle on the prairies was futile without commerce and cash sales. For the settler, the land 'would not be worth the enclosing, and we should see him give up again to the wild common of nature whatever was more than would supply the conveniences of life to be had there for him and his family.'[2] Trade came to the American west over the routes built upon the wilderness, the trails and turnpikes for the waggons, the canals to connect the waterways and the dredgers to aid navigation on the rivers, then the railroads that finally set iron tracks across the fanciful and transcontinental pattern of Thomas Jefferson. In the end, the conquest of the savage was made by the metal edge of American transport and technology.

In *Huckleberry Finn*, Mark Twain supported the escape from civilisation into the wild. But he had already praised the advance of the engine and its navigator onto the great rivers. In his *Life on the Mississippi*, Twain wrote of the riverboat pilot as if he were the Adam of the thundering and churning machine upon the waters. Although there was a science of piloting, it was a curious and wonderful science, and the pilot was the 'only unfettered and entirely independent human being that lived in the earth.' Every man and woman and child had a master, but the Mississippi pilot had none, once the boat was under way on the river. 'His movements were entirely free; he consulted no one, he received commands from nobody ... here was the novelty of a king without a keeper.'[3] Yet the pilot sensed the flow of the mighty stream. He chose to respond to it in his total liberty. Oddly enough, Twain did not seem to notice that the pilot was dependent on the steam-engine. He merely guided its energy through the forces of the currents in a way that was unnatural, a denial of the elements. He was as much a slave of the power of the machine as he was a free spirit navigating the Mississippi.

The same ambiguity about the progress of technology through the wilderness can be found even in the New England pastoral philosophers, Ralph Waldo Emerson and Henry David Thoreau. In a famous entry in his journal, Emerson wrote of the 'whistle of the locomotive in the woods'. Its music was the voice of the nineteenth century, interrogative and prophetic. 'Whew! Whew! Whew!,' it cried. 'How is real estate here in the swamp and wilderness?' Along this passage of iron and steam would come houses and villages and towns. Even Thoreau by Walden Pond emphasised the importance of the Finchburg Railroad bearing travellers from Boston to the countryside. It was the link which related him to society. The iron horse was both reassuring and disturbing, the machine invading the natural order of things. Thoreau did not know how to view progress into the wilderness and ended on a note of hope. 'If all were as it seems, and men made the elements their servants for noble ends!'[4]

The railroads were metropolitan corridors across the continent. Wherever they were built, urban values and markets were carried upon them. Where they were not built, ghost towns faded from hopeful ventures. Locke was proved right. The railroads became the iron arteries of the brief Cattle Kingdom of the plains, the herds driven to the railheads to be transported to the slaughter-houses of Kansas City and Chicago. And they took with them the ultimate barrier that would destroy the great ranches and make the prairies the granary of the world wheat market – the coils of the new invention of barbed wire. These lethal fences kept the savage and the beast from the crops of the farmer on the plains. Hunters and herdsmen were excluded from the growing harvests. More than the rifle and the six-shooter and the railroad, barbed wire won the west from the savage and settled the homesteaders on the far-set corners of Jefferson's grid.

The watermill, the windmill, the magnifying glass, the printing press, and the mechanical clock have been claimed as the chief agents of the 'megamachine' which would overwhelm the earth: Marx himself observed that 'the whole theory of *production and regular motion* was developed' through the clock.[5] But in the west, barbed wire and the railroad were the two main instruments of bringing urban and commercial domination to the wilderness. By the end of the century, the steam engine and world trade could be seen as the necessary, if horrible, mercenaries of the new order. In *The Octopus*, Frank Norris chronicled the war of the ranchers against the railroad trusts, which were 'the leviathan with tentacles of steel clutching into the soil'. The head of this evil metal empire was made to die, suffocated under a cargo of wheat. Yet Norris concluded that the transport system that took the grain to the starving millions of the world was truly progress, even if human greed and cruelty characterised its coming. '*But the* WHEAT *remained*. Untouched, unassailable, undefiled, that mighty world-force, that nourisher of nations, wrapped in Nirvanic calm, indifferent to the human swarm, gigantic, resistless, moved onward in its appointed grooves.'[6]

The next exploitation of this land by the machine would be its greatest, that of the petrol engine and the automobile with the tarmac highway as the link to the world market. And if any images were to depict the conversion of the Russian steppes into a vast expanse of grain, it would be the moving pictures of lines of tractors working the soil and gathering the harvests, where once there had only been hoes and sickles. Marx's vision of man's dominion over a productive earth appeared to be realised. These new engines would also accelerate the logging of the forests with the advent of mechanical saws and pulp-mills, and also lorries for hauling away great trunks and branches. The grasslands and the trees would fall faster in front of the cogs and the gears.

The very beasts fell foul of this remorseless advance of the pistons and the levers and the markets. The herds of longhorns at the railheads were driven onto the cramped cattle-trucks by shouting men wielding steel-spiked clubs. They would arrive at the slaughter-houses starved and waterless, diseased and wounded. After the Civil War, the Cattle Commissioners of New York found 'such an amount of reckless barbarity towards animals and of criminal indifference on the part of many who furnish meat to consumers that one almost wonders why the city has escaped a pestilence.'[7] It was only a foretaste. The era of factory and battery farming, the caging of the beasts and the spreading of parasites through artificial foods, was yet to come.

The technology of the machine seemed to have given to mankind that dominion over the beasts and the plants of the earth, which had been promised to Adam and to Noah in the Book of Genesis. The human race was multiplying and filling the earth, certainly in the United States and in the overseas possessions of the European powers. And with improved communications, the standard of living in the expanding cities of the world was rising. But this new increase was provided by the expanded use of engineering. Abundance, as *People of Plenty* pointed out, is by technology out of environment. New goods, new communications, new sources of energy, new services began to transform the American way of life more than once every generation. 'Whereas our forebears have abandoned only one Europe, they have abandoned several outmoded Americas – frontier America, rural America, the isolated America of the river steamboat and the iron horse.'[8] And at each abandonment, the savage was excluded further from civilisation and its machinery.

The confidence of the Victorian age saw the industrial revolution as the vanguard of civilisation. To Andrew Ure in 1835 in *The Philosophy of Manufacturers*, the new factories of Britain were magnificent edifices superior in value and construction to the pyramids of Europe and the temples of Rome. The growing factory system would become 'the great minister of civilisation to the terraqueous globe, enabling this country, as its heart, to diffuse along with its commerce, the life-blood of science and religion to myriads of people ...' Twenty years later, this theme was taken to extremes by Thomas Ewbank in *The World a Workshop, or, The Physical Relationship of Man to the Earth*. The American author saw the planet designed for a factory and furnished by God, the Great Engineer. Creation had been provided 'for the cultivation and application of chemical and mechanical science as the basis of human development'. It was a religious duty to convert this grand machine shop and laboratory of opportunity into factories and houses and highways and canals. The industrialisation of the wilderness was the divine purpose. Nature was not an ecology, but an economy waiting to be run by humanity.[9]

Victorian imperialism was even more technological and industrial than it was military. The firepower of the fleets and armies of Europe remorselessly destroyed the most elusive of enemies. Artillery, rockets, repeating-rifles and the Maxim-gun saw to the triumph of disciplined expeditionary forces against their forest foes. Garnet Wolseley could even survive his classic advance on Kumasi in the First Ashanti War. He drew up his troops in a square, of which no side could see the other because of the trees and undergrowth in between. He himself sat in a cane chair in the middle, smoking his cigar. The Ashanti infiltrated inside the square, but were killed off by such professional slaughterers as the explorer Stanley, who seems to have personally shot more black men than any other adventurer of his time. However ill-adapted the tactics of the rational mind were to the facts of the forest, fire-power could lead to victory, just as the grid-map could defeat natural geography.

The ranks, wheels and turns of European soldiers that went under the name of drill were, indeed, meant to distinguish a trained from a savage army. When Lord Chelmsford led out his alien white troops and tribal levies against the disciplined Zulu *impis* in 1879, he issued his officers with a pamphlet called *Regulations for Field Forces in South Africa*. In it, he declared that drill was unknown among the Zulus and that, like other savage armies, they had little commissariat or transport. To Lord Chelmsford, the slow length of the British regiments marching in order with their lumbering baggage train was superior to the Zulu *impis*, living off the land on their rations of mealies, capable of quick movement and of concealment in mass, as effective and coherent a weapon as the Roman legions, their battleplan that of Hannibal at

Cannae, a double encirclement by horns of warriors and a mass charge by the centre with its stabbing spears. Such disdain of the Zulu fighting masses led to the disaster at Isandhlwana, where twenty thousand Zulus successfully surprised the British camp from the gullies and crevasses of its site.

If later the laager of chained waggons and the redcoat square on open ground provided the barriers behind which the Gatling guns and seven-pounders could massacre the Zulu charge, Mr Gladstone was correct in accusing Disraeli of allowing the slaughter of tribesmen 'who in defence of their own land offered their naked bodies to the terribly improved artillery and arms of modern science.' The Zulus named their final disaster in front of their king's kraal 'Ocwecweni', meaning 'sheet-iron fort', because they could not even reach the British square as it volleyed out shells and bullets in a storm of metal. 'Into the hordes in front,' an English war correspondent wrote, 'the Gatlings, with their measured volleys, were raining pitiless showers of death ... The Zulus could not get to close quarters simply because of the sheer weight of our fire. The canister tore through them like a harrow through weeds; the rockets ravaged their zigzag paths through the masses.' When the dead were counted, the Zulus had lost fifteen hundred men, the British twelve.[10]

After this triumph, Africa soon gave the first warning of the end of the British Empire. This lay in the inability of its generals and better-equipped armies to cope with the Boer commandos at the beginning of the twentieth century. Kitchener, who finally ended the war, did so geometrically and brutally, drawing and quartering the veldt with barbed wire and block-houses and herding Boer families into concentration camps. 'Like wild animals,' he explained of the Boer rebels, 'they have to be got into enclosures before they can be captured.' Such rigid thinking was to end in the massacres of the fixed trenches of the Great War and finally in the last humiliation of the imperial and mechanistic armies of France and the United States in Vietnam, where all the Western tactics of regrouping natives and bombing villages and defoliation still left the forest guerrillas in charge of their shattered retreats. Those who understand the ground finally control it.

Outside the unequal tally of the battlefield, imperialism itself was an oppression based on new methods and annual statistics. To the villager in the back country who could hunt or grow enough to satisfy his wants, only the imposition of taxes could force him into regular work. The necessity of money related him to the Western concept of arithmetic and time. To collect that money so essential for the colonial governors and their budgets, communications had to be improved. And nothing was more dominant to the savage eye than the authority of the armed steamship cutting through the meandering waterways with its iron prow, or the gash that the railway ruled through the sprawl of the trees, or the weight of the bridges that bore the new roads between market and town. The Romans had once imposed a quadrilateral of straight military roads on the wooded face of England to separate and govern the tribes. Now the British also used the pattern of road and rail to divide the forests of Africa, which had not yet known any human master in history.

After the engineer and the soldier came the administrator and the tax-gatherer. The ledger became an even mightier weapon of oppression than the musket or the Mauser. Where Stanley had hacked out stations in the Congo bush for Leopold, King of the Belgians – the seeming advance posts of morality and progress – he discovered in the end that they were the notorious barracks of colonial exploitation. By the use of cannibal mercenaries and greedy company employees, Leopold destroyed the village culture of the Congo and turned its people into serfs in search of raw materials. Each village was assigned a monthly quota of rubber, ivory and food for the soldiers; the penalty for producing too little was the flogging of the men and the chaining of the women; resistance met with utter destruction. The soldiers moved on from ruined village to untapped village, while the output of rubber soared and Leopold grew richer. He had promised 'moral and material regeneration' to his fief. He gave it ruthless rule and an export surplus. The soldiers had to produce a severed right hand for each cartridge expended on their mission of keeping order and getting rubber. As Herbert Samuel asked in the House of Commons: 'If the administration of the Congo state is civilisation, then what was barbarism?'[11]

Reluctantly, tardily, after Sir Roger Casement had reported on the horrors of the Congo, the Belgian parliament sent out a judicial commission to investigate. The judges whitewashed the atrocities, but they did admit that Casement was right in his general thesis. The official who produced the most rubber received the quickest promotion. His agents were 'admittedly savage soldiers', and their assaults had produced the terror in the Congo. The sap of the rubber tree was the root of all evil. The commission knew its brief well enough to justify forced labour as the only method of taxation by which 'the population can be reclaimed from its natural state of Barbarism', although it did recommend that taxes should be limited to the strict necessities of the state. And so, through the system of taxation, the trees of the rain forest became the factories as well as the refuge of the natives.

The savagery of any society is usually increased by the exactions forced upon it, and the Congo was no exception. Where dog eats dog, man has to feed on man. While cannibalism was used ritually in tribal Africa and served to counter a protein deficiency, it was rare until the slave-trade depopulated settled villages and devalued human life. Stanley on his first expedition from East Africa to the Congo unwittingly opened up the way for the Arab slave-traders; by the time of his reverse journey, he found devastation and cannibalism widespread in the rain-forests.

Nothing more shows the degradation caused by the European pressure for slaves or ivory or palm oil than the degeneration of the proud city-empire of Benin. When the Europeans had arrived in the sixteenth and seventeenth centuries, they had found a great city with a main street, 'seven or eight times broader than the Warmoes Street in Amsterdam' and houses bordering it 'in good order, one close and even with the other, as the houses in Holland stand.'[12] The government and the people were easy in their trade with Europeans, and their brasswork was at its height of excellence. Yet the endless wars necessary to collect slaves for the trade in European goods led to Benin's decline into disorder and ritual sacrifice. By the end of the Victorian period, the massacres at Benin stank in the nostrils of the new Nigerian administration, and as the threat of a British punitive expedition increased, so the power of the doomed fetish priests ended in an orgy of blood lust. When the city finally fell in 1897 under the iron attack of the Maxim guns and the shrapnel of the shells and rockets, it was found to be a city of skulls worse than Tenochtitlan. 'Every person who was able,' an observer reported, 'indulged in human sacrifice, and those who could not, sacrificed some animal and left the remains ... the whole town seemed some huge pest-house.'[13]

So commercial pressures bred savagery, which in turn confirmed moralistic efforts to civilise tribalism. Victorian missionaries rarely considered that some of the evils which they came to extirpate were the results of European pressures upon pastoral societies. The interminable thefts of trade goods by aborigines were seen as a basic sin of character rather than as a temptation in the face of the glittering unknown. The early acceptance of strange habits by the explorers changed to condemnation by the missionaries. Bruce in Abyssinia merely described the custom of hacking live steaks from a bull; its bellowing was a signal for everyone to sit down at table and eat the raw meat. Yet for Livingstone, the eating habits of the bushmen put them on a level with the beasts. In his diaries he noted that the primitives ate exactly like the animals and birds; he wrote equally of the Bushman and the frog called Matlametlo.[14] He did not admire these expert survivors of the desert for learning to live on what their close association with nature had taught them. Missionary bias made him fail to distinguish an alternative culture from an innate savagery.

Where local custom did not fit the rule, it was wrong. Victorian pioneers still aimed to judge, not to accept. And the popular literature of the time predetermined in Europe the general perception of preliterate peoples. Where *Robinson Crusoe* had presented a sympathetic portrait of the savage Friday, *The Coral Island* gave a degraded and simplistic description of Polynesian societies. R.M. Ballantyne's 'classic' novel of 1858 featured shipwrecked English boys as the observers of a tribal war. Their childish morality was superior to local anarchy. Hidden behind bushes, the young white hero watched the battle. 'There was no science displayed. The two bodies of savages rushed headlong upon each other and engaged in a general mêlée.' They looked like incarnate friends, 'more like demons than human beings.' The winning chief was the biggest and most bloodthirsty of them all. The law of the islanders was the apparent law of the jungle. 'The island is inhabited by thorough-going, out-and-out cannibals,' the hero pronounced, 'whose principal law is "Might is Right and the weakest goes to the wall".' But even he was shocked by the wanton slaughter of 'a bloody and deceitful tribe' of savages, when a pirate and trading vessel turned Long Tom on them and the big brass gun cut a lane through the living men with a withering shower of grapeshot. Such execution horrified the young hero, since the pirate captain also favoured the missionaries 'because they are useful to him and can tame the savages better than anyone else!'[15]

Such a crude exposition of the virtues of Christianity in making foreign cultures amenable to exploitation by trade, or of mutilating the resistance of the savage by the war machines of Europe, was the popular view of the justice of imperialism. If it was not always right, it was necessary. The progress of civilisation was inevitable. This was the directive of the new science and industry in making a workshop of the world, as well as Darwin's precept of the survival of the fittest as applied to human societies. Only when trading in far places began to give way to modern anthropology and to tourism would other perceptions of distant peoples begin to emerge.

8
THE CULT OF THE SAVAGE

THE CULT OF the savage increased with the fashion for tourism. Those who deliberately chose to inflict travel upon themselves in the centuries of bad roads and hotels before the coming of the railways set out on their journeys knowing that they would be surprised as well as informed. They travelled hopefully to arrive at the terrors of the unknown as much as the marvels. To such sybarites as Horace Walpole, even Sussex was too distant a visit from London. 'The whole country has a Saxon air, and the inhabitants are savage ... Sussex is a great damper of curiosities.'[1] Those whose curiosity still remained keen journeyed on, seeking the excesses of nature and of man, particularly in that age of heightened perception which is called the time of the Gothic or Romantic imagination.

There is no question that, before the middle of the eighteenth century, most travellers looked for antiquities and royal palaces, new building and farming methods; nostalgia and social progress interested them far more than the wild parts of the country or its inhabitants. East Anglia was praised for its arable land and flatness, while the county of Westmoreland seemed to Defoe 'eminent only for being the wildest, most barren and frightful of any that I have passed over in *England*'.[2] When faced with the Alps, the essayist Addison did admit to a sort of agreeable shuddering, but he finally condemned 'this most misshapen scenery.' Yet when the classical admiration for the order put by man on nature gave way to the romantic fashion for the terrible and the sublime, then the traveller sought and found the savage, both in the new industrial processes that were brutalising the land of England and in the ancient upheavals of geology that had cracked open the gorges and fissures of the mountains.

The awfulness of Ferreday's new ironworks at Bilston found one traveller in 1810 conjuring up a vision of the deformed Cyclops forging thunder for Jupiter. The appearance of the workmen, covered with dirt and working madly to keep pace with the violence of the machinery, filled the visitor with terror. 'A sight of this concussion of wind, fire and water' driving on its frantic slaves at the furnaces, exceeded all ancient and modern descriptions just as much as reality exceeded imagination.[3] Yet the traveller had brought his nightmare of a classical inferno to Bilston with him; he had sought out the ironworks, not for profit or labour, but for the pleasure of the shock and for curiosity. His wild view of the puddlers and the smelters was as artificial and consciously romantic as the Gothic writer Mrs Radcliffe's description of Borrowdale Gorge. 'Dark rocks yawn at its entrance, terrific as the wildness of a maniac; and disclose a narrow pass, running up between mountains of granite, that are shook into almost every possible form of horror.'[4] To the works of nature and of men, the travellers after the Age of Reason brought their new perceptions and opera glasses, seeking out the frenzied and the exaggerated and the strange.[5]

Those who seek shall find. So it was promised in the Bible, and in the Baedeker of later days. The Byronic cult of the savage induced a worldwide temptation in the rich and the secure, to search out the solitary and the strange. This passion for the primitive was enhanced on the east coast of the United States by the recent removal of the Indians, which created a nostalgia for them, now that they were more of a memory than a threat. A deliberate flight from the values of the city and the town took place, most apparent among the Transcendentalists of Boston. Emerson, who had made the city with its finite lines and calculations the mere home of understanding, declared that the country with its varieties and irregularities was the home of reason, and indeed wisdom. On a walk across a safe town common, he felt a wild poetic delight. 'I am glad to the brink of fear. In the woods, too, a man casts off his years, as the snake his slough, and at what period soever of life is always a child. In the woods is perpetual youth.'[6] So also Thoreau hymned the virtues of the forest, withdrawing just as far west as Walden Pond, while telling the young to turn their backs on Europe and move as far as Oregon. The later historian of *The Oregon Trail*, the Brahmin Francis Parkman, fell in love with the wilderness while still cloistered safely in Boston. As he confessed in his journals, even while he was a youth, 'his thoughts were always in the forest, whose features possessed his waking and sleeping dreams, filling him with vague cravings impossible to satisfy.'[7]

Jung would later have a word for these yearnings towards wild behaviour and wandering. He would call it an obsession – something as natural in urban man as it was in the Indian. The problem was that, while an Indian could say he was plagued by a ghost of the past and be understood for such an admission, a civilised decadent such as Francis Parkman could only admit to a trick of the imagination or to a nightmare. Yet, as Jung pointed out, the primitive phenomenon of obsession had not vanished; it was only interpreted in a different and more obnoxious way.[8]

While not yet recognising that the gratification of an occasional savage impulse might be desirable from time to time, even in the most urbane of men, the Transcendentalists did admit to the special gifts of those who had remained in the forests, close to nature. Thoreau himself saw the destructive edge of the process of trying to educate primitive men. 'We talk of civilising the Indian,' he wrote, 'but that is not the name for his improvement.' The wary independence and aloofness of the Indians in their forest lives allowed them a rare and peculiar contact with nature. In that lay their genius, and it had been lost to the dwellers of the towns and the cities. There could be no quick Wordsworthian trip to the trees to achieve on a ramble the wisdom of nature. Wordsworth might claim:

> One impulse from a vernal wood
> May teach you more of man,
>   Of moral evil and of good,
> Than all the sages can.

Yet such an easy illumination was simply untrue. Only those who lived long in the woods could have such perceptions. For their nature then recalled their surroundings and allowed a communication between them.

Many European and American commentators in Victorian times harked back to the claims of Tacitus about the virtue of the ancient Germans of the forest. If the New World was really to breed a new form of man, he must be the product of his experiences in the woods, now largely lopped and burned in Europe. The first major American historian, George Bancroft, had even claimed a backwoods life for George Washington to prove his special virtue as the Founding President, while the ambiguous interpretations of the natures of the men of the forest given by Fenimore Cooper and Hawthorne did not prevent the myth of the goodness of the backwoods from remaining a dominant theme in American literature and history throughout the nineteenth century, once the Indians had been pushed out of the trees into the exposure of the prairies.[9] The useful and sheltering woods became the materials and the complacent mythology of the mid-western small town, which saw in them a resource and a protector rather than a menace. The very savage was lauded by Walt Whitman in his *Song of Myself*:

THE CULT OF THE SAVAGE

> The friendly and flowing savage, who is he?
> Is he waiting for civilisation, or past it and mastering it?
> Is he some Southwesterner rais'd out-doors? is he Kanadian?
> Is he from Mississippi country? Iowa, Oregon, California?
> The mountains? prairie-life, bush-life? or sailor from the sea?
> Wherever he goes men and women accept him and desire him.
> They desire he should like them, touch them, speak to them, stay with them.
> Behaviour lawless as snow-flakes, words simple as grass ...

If Whitman did not answer his question of whether the savage was expecting civilisation or already past it and mastering it, he did make out that the untutored and provincial American was already the most wanted man on earth, the potential maker of a New World.

The most famous theory about the American wilderness, 'The Significance of the Frontier in American History', was delivered as a paper in 1893 by Frederick Jackson Turner. He tried to prove that the idealisation of the pioneers of the forest and the wilderness was part of an actual historical process in the United States. He defined the frontier as 'the meeting point between savagery and civilisation.' Its site was where the cleared field met the trees or the mountain. Thus the frontier was a moving contact; it advanced with the farm and receded with the shrinking area of wilderness. The economic fact of free land attracted and bred a liberty of the human spirit. American democracy, as Turner declared twenty years later, 'came stark and strong and full of life out of the American forest, and it gained new strength each time it touched a new frontier.'[10] Its ancestor was the free democracy of the German woods praised by Tacitus, its heir the political liberty of the American people.

Yet as Turner spoke, the American frontier had officially ended. There was no more free land to act as a safety valve for social danger. A new frontier was being drawn between the virtues of the forests and the small towns, and between the alien values of the cities advancing upon the cleared countryside in a second inescapable invasion. Metropolitan tourists might still search for the good life in the backwoods, but they would often discover that the city had broadcast its influence over the land before them, and that their visit chased away the peace which they had hoped to find. If Turner had used the word 'savage' instead of 'frontier' to describe the engagement between early societies and geographical areas and urban pioneers, he would have had a term of more resonance and expectation.

* * *

The sylvan virtues hymned by Whitman and Turner seemed even more desirable when contrasted with the evils of the expanding slums of the coasts and of the communications centres in the interior. For the industrial revolution in the Western world had created urban masses grouped round the factory cities and swollen capitals of the great powers. These new slum populations became the terror of the literate and godly and growing middle classes, which had erected a high standard of living on the wealth produced by the factories. The proletariat in the ghettoes seemed as fearsome now to the clerk in his villa as the Iroquois in the forest had once seemed to the Puritan in his frame house. The wandering millions of tramps and hobos and casual labourers of America in the 1890s were as menacing to the villages and farms as the sturdy beggars had been to Elizabethan England, and they invited the same repression. The forests and the aborigines may largely have been contained or destroyed in late Victorian Europe and the United States; but a new wilderness had grown up of brick and street, of hovel and back-to-back and shanty and closemouth, an urban jungle in the labyrinth of foul streets at the very doors of the rulers and the privileged. No city walls nor stockades now kept casual violence or mob attack from the man of property. The new industrial cities contained their enemies within their boundaries. The slum came to be called a jungle, and the term 'the savage' so lost its meaning that it was applied to the thug of the alleys, the street arab and the apache of the Paris gutters rather than to the native of the woods and the wilderness, now gathering qualities like moss as the cities forgot about him.

Social Darwinism had brought about this transfer of meaning from forest to slum. Darwin himself had recognised at the end of his life that his theories might be applicable to the social situation of the nineteenth century. With Marx as one enthusiastic disciple, seeking to use the conflicts of nature in the service of the dialectic of revolution, and with Herbert Spencer as another disciple, seeking to justify the harshness of *laissez-faire* capitalism by an appeal to the necessary survival of the fittest, Darwin became an apostle to both of the leading economic theories of the twentieth century, although he had never claimed to be more than a biologist. The use of his concepts of natural selection by social philosophers started a fashion – now more than a century old – for seeking to explain the behaviour of urban men by the repressed and inherited instincts of his savage past as well as by the observed habits of wild beasts. Kropotkin, indeed, asserted the opposite of Darwin, that natural history demonstrated the mutual aid necessary to a surviving species rather than the struggle between the weak and the strong. In one sense, the old argument between the bestial savage of Sepúlveda and Hobbes as opposed to the good savage of Las Casas and Locke had transferred its dialogue into biology. Both conservatives and revolutionaries preferred to use the cruelty of Social Darwinism as a justification for their indifference to change or for their call to revolt. Only the gradualist reformers thought that co-operation might be a better principle for the improvement of the species than natural selection.

Through the writings of the realist school of American novelists of the late nineteenth century, particularly Howells and Crane and Norris and Dreiser, the idea of the city as a vicious jungle began to push its way into literature. The metropolis no longer seemed a beacon of civilisation or even an inferno of delights to the brutish countryside; it appeared as a battlefield between the privileged and the degraded with a good chance that the savage elements might triumph. As Dreiser wrote in *Sister Carrie* in 1900; 'Our civilisation is still in a middle stage, scarcely beast, in that it is no longer wholly guided by instinct; scarcely human, in that it is not yet wholly guided by reason. On the tiger no responsibility rests.'[11] Yet the tiger was to other American socialist writers such as Jack London the jaws of the slum itself, which the city had let inside its gates to devour its own young.[12]

As both a dedicated Spencerian and a Marxist, Jack London was the first popular author to equate the wilderness with the ghetto as the basis for a book. His experience of slum life in Oakland outside San Francisco, followed by his struggle for survival in the Klondike gold rush and Alaska, made him translate savagery directly from the snows to the streets. Always a man of strong prejudices and received social theories, he had judged the slums of New York and Europe before he saw them. His one effort at sociology, *The People of the Abyss*, contained more indignation than information, based as it was on seven weeks spent in the slums of the East End of London after the imperial coronation of Edward the Seventh. Before he reached England, he had already arrived at his conclusions about the poor workers whom he would meet. 'That's all they are – beasts –' he wrote on the boat to England, 'if they are anything like the slum people of New York.'[13]

Expecting to see an urban wilderness, Jack London discovered a monotonous slum where even the struggle for survival hardly stirred the general apathy and hopelessness of an enfeebled stock. Yet on the way to the docks late at night, the American visitor saw the wilder view of the East End which Darwinism had already formed in his mind. To him, the scene 'was a menagerie of garmented bipeds that looked something like humans and more like beasts, and to complete the picture, brass-buttoned keepers kept order among them when they snarled too fiercely.' These small and misshapen gutter-wolves with the paws of gorillas were 'a new species, a breed of city savages. The streets and houses, alleys and courts, are their hunting ground. As valley and mountain are to the natural savage, street and building are valley and mountain to them. The slum is their jungle, and they live and prey in the jungle.'[14] Woe to the dear soft people of the West End, when the soldiers of Empire were fighting abroad, and these slum beasts would crawl out of their dens and lairs to attack the aristocrats. It would be another French Revolution.

Jack London came to the final conclusion that the Innuit Indians of Alaska with their subsistence hunting economy were better off than the average English worker with all his imperial and industrial power. Huxley was correct in preferring the life of the elementary savage to the life of the poor in Christian London. The job of civilisation should be to better the lot of the citizen, not to worsen it. Economic mismanagement in a rich country had reborn the savage in the city streets. Socialism was now the last hope of saving the urban poor from blight and early death. Although the power of producing food and goods had increased a hundredfold, 'through mismanagement the men of civilisation live worse than the beasts, and have less to eat and wear and protect them from the elements than the savage Innuit in a frigid climate who lives today as he lived in the Stone Age ten thousand years ago.'[15]

So Jack London developed the case that the evils of industrialism had reduced men to the bestiality and the slavery to necessity from which hundreds of years of village and urban life had sought to rescue them. His theme was taken up even more strongly by another socialist writer, Upton Sinclair, who rose to fame three years later in 1905 by equating the steaming ghetto with the equatorial forest. *The Jungle* was the title of Sinclair's tract of a novel, which was set in the slums surrounding the slaughterhouses and canneries of Chicago. This bloody battleground of life was presented crudely and effectively, far from the veiled decencies of Dreiser's Chicago. To Sinclair the very smell of the city was strong and rancid and sensual. The slaughterhouses were rapids of death which gorged on transported and diseased cattle and hogs, then spewed them out as the canned mincemeat of civilisation. Trapped in this mass-murdering mechanism, men and women and children fought for filthy jobs and were doomed to lose the struggle for existence either quickly or slowly – only the loss was certain. Outsiders would visit them at their terrible toil in the stink of the fertiliser plant or at the sausage machine, and they would be regarded as if they were 'some wild beast in a menagerie.'

In this waste of mean streets, the old primeval fears of life rose screaming in the throats of the recent immigrants from Europe. They may have sailed across the Atlantic to escape the dull forest surrounding their village at home; but they now found themselves in a worse hovel surrounded by the urban jungle. Jurgis, the hero of *The Jungle*, began by believing in the survival of the fittest; but he ended by looking for his own survival at any price. For in Chicago, 'human beings writhed and fought and fell upon each other like wolves in a pit ... Into this wild-beast tangle these men had been born without their consent, they had taken part in it because they could not help it.' So Jurgis fought while his strength lasted and lost when his strength went. 'He was like a wounded animal in the forest; he was forced to compete with his enemies upon unequal terms. There would be no consideration for him because of his weakness.'[16]

So Upton Sinclair presented the city as *The Jungle*. The book ended with the hope of socialism to come, with strength given to the masses through unions and co-operative farms and state factories, the mutual aid advocated by Kropotkin in revolt from the economic struggle excused by the social Darwinians. Although Sinclair's writing stimulated the passage of laws to control the stockyards and the quality of American food, his message of co-operation rather than competition found little support. As Oscar Handlin has suggested in his remarkable essay, 'The Horror', there was too much fear and mobility in society at that time for an acceptance of new ideas or a trial of the brotherhood of man. There was the constant danger of loss of status, as the small town and the family farm declined in importance and the factories grew larger with their devouring chimneys. 'It was all very well to let the mind wander in the reveries of local-colour literature about the bygone past. But no American could disregard the growing armies of the proletariat, of the hired labourers and the tenant farmers, of the millions of tramps – all existing in the brutish misery that was the penalty of their failure.' The risks of falling into the abyss were as great as the chances of rising to the rich heights promised by the perennial optimism of an Horatio Alger.

This sense of threat, of an alienation from a promised land, was not only an American phenomenon at the turn of the century. In the great cities of central Europe a massive and heterogeneous population was collecting. This new proletariat and pauperised *petite bourgeoisie* were cut off in the slums from their traditions and the certainties of usual family life. They feared the loss of their inherited personalities and 'recoiling in horror, sought an antagonist to hate in order to discover their own identity.'[17] So doing, it was their new horror that their prejudice could become the weapon in the hand of the rising totalitarian state, which would pervert this despair of the dispossessed into a hatred of a named dispossessor, a chosen enemy who might attract all the fears and loathing, all the suppressed ferocity and need for release which are gathered under the name of 'the savage' in man.

The First World War was the condition for the destruction of the old orderings of European Life. As Hannah Arendt has written, the days before and the days after the war are separated not like the end of an old and the beginning of a new period, but like the day before and the day after an explosion, which began a chain reaction of explosions that the world is still trying to survive.[18] During the four years of the war, tens of millions of men learned the mindless obedience and mass comradeship needed to withstand the static campaigns and massacres in the trenches, while hundreds of millions of women and the young and the old became resigned to the bureaucratic incapability to share out insufficient food behind the lines.

The aftermath of the war was worse: rampant inflation, which destroyed those bourgeois virtues built upon the value of money; mass unemployment, which made the dignity of labour seem like an obscenity; millions of stateless people, the victims of civil wars and the new national states of central Europe, who lost their legal rights with the loss of their homelands, to become the wandering savages of the 1920s in a Europe preparing itself for totalitarianism and the persecution of minorities. For Arendt has also argued persuasively that the stateless may be well-educated, but they are basically savages, since all they have is their inalienable human rights which any savage has from the fact of being human. What they needed after the Great War was national rather than natural rights, which might protect them by law from the harrying and the concentration camps to come. If it was the tragedy of savage man briefly to inhabit an unchanged natural world, upon which he left no mark or trace, then the stateless and shifting millions, able to contribute next to nothing to their temporary abodes as they passed through the frontiers, were the evidence of a possible regression from civilisation.[19]

The First World War had another major effect. The easy belief in mechanical progress of the late Victorians became fouled with the butcheries and horrors created by that industrialism which had produced the conscript armies and had sacrificed them on the barbed wire and machine-guns and poison gas and massed artillery of the new technology of war. The parallel rise of psychiatry through the influence of the pupils of Freud and Jung also made men increase their questioning about the motives behind patriotism and self-sacrifice for the call of duty. The myths conjured up by governments to mass men for war, and the monsters created to make soldiers kill their enemies, were revealed as the mere disordering of a few men's imaginations broadcast over the minds of many people. What were truly fearsome were the new machines of death, which could slay like David in his tens of thousands. The unknown and the hostile, however, became less frightening as men began to recognise that they need have no fear but fear itself, and that the visible savage who frightened them was often merely the invisible savage who lived within them and needed to emerge.

This summary of the interaction of some of the forces moving the Western World after the Great War concentrates upon those elements which were to blur the certain Victorian categories between right and wrong, civilised and uncivilised, law and aggression. The habit of military obedience, the acceptance of short rations, the apathy of mass unemployment, the worthlessness of paper money, all these contributed to the breakdown of that sense of individual pride which had enabled the white man to dominate lesser breeds in the European colonies of the world, without questioning the assumptions of his role. When a feeling of powerlessness in the face of the advance of the machine and the metropolis was added to a method of seeking the hidden motive behind every direct action, then the old sureties of Victorian progress became merely the crutches of a bourgeoisie that had once ruled, but now feared it could not even stand pat in post-war society, let alone serve as the vanguard for more economic progress.

As the rise of Fascism showed, there was also a positive regression towards barbaric values. In a sense, the last refuge of sophistication is nihilism; total decadence looks to total destruction to purify itself. The savage Boadicea burning London and the effete Nero burning Rome reached the same conclusion. Without the devastation of the old, nothing new could be created. The condition of the rise of the phoenix was the making of its womb of ashes. This spirit of Herostratus was widespread among the pre-war intellectuals and decadents; it made them leap towards the promise of an apocalypse or a *Götterdämmerung*, in order to will upon their civilised values the barbaric annihilation of total war. German, Italian, French and British memoirs of the time all invoked the same yielding to relief that the ordered world was about to be shaken apart, and that 'ye Barbarians, Scythians, Negroes, Indians' would now trample upon the interests of civilised society that had slowly accrued throughout the centuries.[20] As the founder of Futurism, Marinetti, declared, 'We will glorify war, the only true hygiene of the world.'

Hitler and the Nazi party harnessed both of these positive and negative forces during the return of Germany to barbarism under the rule of the Third Reich. The Führer exploited the submergence of the individual in the mob, calling it the rise of the folk consciousness, when it was actually the reversion to the Hun horde, careless of life in its mass pursuit of conquest. He chose enemy after enemy to give the Germans a sense of superiority – Jew and gypsy and Slav – until he named the final enemy, Germany itself, unfit for the glories of the Third Reich, therefore doomed for a thousand years to the total destruction which it had wished on others. He encouraged the private savagery of man against man, the torture and bestiality used by the SA and SS against their victims, the degradation of whole populations whose land he coveted and whose death he ordered. Worst of all, he seemed able to prove his terminal case, that 'inferior' species of men could be reduced in the filth and anonymity of concentration camps to behaviour not better than animals, which they were held to resemble more than they resembled Germans. As Arendt bitterly observed: 'Actually the experience of the concentration camps does show that human beings can be transformed into specimens of the human beast, and that man's "nature" is only "human" insofar as it opens up to man the possibility of becoming something highly unnatural, that is, a man.' [21]

If this Nazi atavism was finally defeated at the cost of tens of millions of lives, it proved that the powers of barbarism were still most powerful and could be exploited in one of the more civilised nations of the world. The perversity in urban men, their wish for self-destruction as well as their root obsession for the forgotten and wild life of the hunting forest, all were the strings waiting to be pulled by the demagogue, playing puppet-master. The cult of the savage, which early tourism and the Romantic movement had initiated as a holiday and a release for urban man, had turned into a bourgeois fear of the savage within the city slum, and finally into a whole national philosophy of savagery, in which each individual and interest and class would surrender its will into the hands of the Führer, who would then direct the folk into actions outside the limits of mere morality.

Yet the whole of the Western world in the 1930s was not yet the Third Reich. As Fascism revealed its regressive features year by year, as its love of violence and mob ritual offended that diminishing band of intellectuals who still believed passionately in human rights and did not join in the kow-tow of *la trahison des clercs*, so a fresh group of anthropologists and writers began to examine the last of the primitive peoples in an effort to find out whether they possessed secrets of living that might help to mitigate the madness in modern Europe.

On the coral archipelago of the Trobriand Islands of northeastern New Guinea, Bronislaw Malinowski found his society of savages, which might confound the assumptions of the emergent psychoanalysts and totalitarians. He approached his field work with certain assumptions and came to certain conclusions, that the primitive mind was not so different from the civilised mind, that it did have lucid intervals in a world threatened by irrational and supernatural dangers, and that it would follow its own self-interest where necessary against the dictates of clan custom. Before 1926, when Malinowski's works on crime and sex and repression in what he ironically termed 'Savage Society' began to be published, anthropology had already developed from collecting the quaint habits of primitive peoples, in order to prove their inferiority and lawlessness, into a total effort to record their tribal lives, in order to prove that they were wholly bound by totem and taboo and group – in fact, that they were among the more law-abiding peoples on earth.

Both the anarchic savage and the primitive slave to custom were as convenient to contemporary thought as the devilish and the noble savage had once been to Hobbes and Locke. To Fascists, the lawless savage seemed to be a racial inferior analogous to a gypsy. To Communists, his total immersion in clan or group life was a primitive version of that brotherhood of man which had been perverted by the rise of capitalism and which would come again with the withering away of the state. Malinowski's aim and achievement was to show that his Trobriand Islanders obeyed most of their laws and customs spontaneously, evading a few restrictions just as city people did. Their reasons for general obedience and sociability were as inherited, complex and psychologically necessary as those of any Western man – it was certainly not the threat of coercion that made them behave well. Malinowski's savages were not ruled by mood, passion or accidents, but by tradition and order. Their difference merely lay in their punishments when they broke the law – ostracism or public shame forcing them to suicide were the ultimate penalties of their society.

Malinowski further showed that Freudian assumptions about a universal Oedipus Complex, for instance, were ridiculous in the matriarchal Trobriand society, in which every child had two fathers, his mother's brother and his own father, who was also his nurse and companion. Altogether Malinowski stressed that there was no 'group mind' or 'collective unconscious' beloved of totalitarian and monist thinkers, only a complex relationship between biological, psychological and sociological factors which made savages evolve one way of living and Europeans another.[22]

In 1927, two other important anthropological works were published, Franz Boas's *Primitive Art* and the work of his pupil Paul Radin, *Primitive Man as Philosopher*. Like Malinowski, both scholars reacted from the old anthropological view that a 'savage' culture was an early and debased form of a civilised culture; to them, it was a different style of living produced by men of fundamentally the same intellect as civilised men, but brought up in a different manner with different materials at hand. They also reacted from the idea of primitive man as anarchic or bound by custom. They found him as individualistic and responsible to society for his own thoughts and actions as any civilised man. Boas even declared of the Andaman Islanders, still at the primary 'savage' stage of spending the bulk of their time in collecting food, that some of them were of a philosophic cast of mind and might reinterpret the current tribal myths and tales of nature.

What the new school of anthropologists found was that nearly every society, whether advanced or primitive, broke down into a majority that desired customary ways, and an intellectual elite which indulged in speculation and scepticism. As Boas declared, in the narrow field of art that is characteristic of every people, the enjoyment of beauty is quite the same in primitive and civilised societies, and its effect is equally intense among the few, slight among the many. If primitive people are more ready to abandon themselves to exaltation through art, civilised people are able to perceive beauty in more complex and catholic ways. Finally, only the quality of his experience, not of his mental powers, distinguishes the savage from the urbane art critic. As Malraux was to put the matter so clearly in *The Voices of Silence*: 'Just as Athens discovered the artistic value of the perfect breast, so the barbarian artist discovered the hawk's beak, the claw, the horn, the fang, the skull, the death's head grin.'[23]

Boas particularly looked to rhythm and dance as something which was accentuated in a tribal group. He saw its formal patterns as an aesthetic of motion. His comments had developed a long way from those of Father Cavazzi, who had observed the Congolese dancing in 1687 and had found that they did not perform a minuet. 'Dancing among these barbarians,' the father had complained, 'having no motive in the virtuous talent of displaying the movement of the body, or the agility of the feet, aims only at the vicious satisfaction of a libidinous appetite.'[24]

Early black music and dancing in the United States were judged equally harshly by their white critics, as a shuffle and a stomp and a shout, neither a folk song nor a hymn nor a brass band, an unholy trinity of discord. Yet out of this unlikely combination evolved ragtime and then jazz, contributing to America its first indigenous music, however much it was originally dismissed as lecherous and savage in the drawing-rooms of Boston and New York before the First World War. Yet the success of jazz in the 1920s proved the power of the new relativist anthropology as well as the force of the new music; all vigorous cultures were now thought to have something to offer, whatever their source. Jazz was, indeed, the first national music to grow out of the rural shack and the black slum. It developed in the Mississippi Valley, spontaneous and antagonistic to all European music, a curious coda on Frederick Jackson Turner's theory of the frontier. For if a new man was not born on the old frontier, a new music certainly was. It spread through America and the world until it had conquered even the concert hall within seventy years and had imposed its formal abandon, its melodic liberties and its ferocious beat within the very drawing-rooms of the privileged, whose lives tried to exclude what their gramophones introduced.

This effort to rediscover the strength and virtues of primitive rhythms and styles of life was foredoomed. Urban Western man could only visit these alien scenes of apparent release and social ease and daylong pleasure on tropical islands as a watcher from outside the bars, no more included than the Georgian tourist had been in the hellish toil of the ironworks at Bilston. The few anthropologists who did spend years of field work living with their chosen tribes still returned to the universities and the cities to publicise their results. Total refugees from civilised values were rare, and when they were reintroduced to these lost temptations, they usually fell victim, like Aldous Huxley's Savage in *Brave New World*. As in the days of Columbus and Raleigh, the savage in paradise became the lost dream of the sophisticated few, who regretted an innocence which they had never had to lose. He was worth a visit and a book, not worth the tattoos of a ceremony of initiation.

D. H. Lawrence, who stood as much with the totalitarian vision of 'rare consciousness' in *The Plumed Serpent* as he did with the values of democracy in his own life of individualism, took pleasure in pointing out the fallacy of urban man's yearning for the primitive life. In his comments on the novels of Fenimore Cooper, he asked whether there was not a residual guilt in the new cult of the savage. It was true that the land had been cleared in America and the Indians had been decimated by a previous generation. 'But is a dead savage nought? Can you make a land virgin by killing off its aborigines?'

In Lawrence's opinion, the Indian, whether alive or dead, would continue to hate or haunt the white man. He was dispossessed and unforgiving; therefore he must appear in the eyes of most white people 'subtly and unremittingly diabolic.' The whitewash of the new social anthropology was useless. It was the hypocrisy of the few slapped upon the mutual and natural loathing of the many. So Lawrence wrote in his terrible summing-up of the inevitable permanent misunderstanding between the civilised and the savage, between the spoiler and the despoiled, between the recorder and the victims of his recordings:

> The desire to extirpate the Indian. And the contradictory desire to glorify him. Both are rampant still today.
>
> The bulk of the white people who live in contact with the Indian today would like to see this Red brother exterminated; not only for the sake of grabbing his land, but because of the silent, invisible, but deadly hostility between the spirit of the two races. The minority of whites intellectualise the Red Man and laud him to the skies. But this minority of whites is mostly a high-brow minority with a big grouch against its own whiteness. So there you are.[25]

9
THE RESPECTABLE SAVAGE

THE SECOND WORLD WAR did not induce the psychic shock of the First, that fissure in memory which dated most human experience to a time before and a time after its outbreak. It merely accelerated the erosion of the old imperial powers and accentuated the divisions of human societies which had always existed, but had been obscured by the ideology of imperialism. The Victorian belief in progress, the Kipling doctrine of the white man's burden, and the old anthropology of Tylor's *Primitive Culture* had all agreed that the duty of civilisation was to educate the savage until he became a copy of his masters. After 1914, doubts about these beliefs had even spread among the imperial administrators themselves, so that they had deliberately left some savage peoples in a condition of backwardness for fear of corrupting them by civilisation. Malinowski himself had wickedly suggested that the new anthropology should be encouraged by the colonial powers, because a practical knowledge of the behaviour of primitive tribes would help the white man to govern, exploit and 'improve' the natives, if he wished, with less pernicious results to victims.[1]

One brilliant novel by Joyce Cary, *Mister Johnson*, showed the ambiguous effect of bringing communications and civilisation to the African. Mister Johnson himself is a clerk in the colonial government, a fantasist who loses his own culture for a dream of an England that exploits him. He ruins himself by marrying a girl from the bush and by trying to turn her into his illusion of a British lady. His brief triumph is to fall into his master Rudbeck's obsession about cutting a road through the forest to bring progress to the old walled city of Fada. After phenomenal labours, the road is created. 'On every side the enormous columns of the trees stand dusty and motionless as stone, under the dark roof of foliage. The narrow crack in that roof, which lets in a strip of light sky above the brown strip of road, is like a single knife cut, already closing. Branches reach across overhead; at a little distance the edges seem to join.' Yet the road exists. Motors find Fada, followed by trade and the influences of the city and the coast. Rudbeck, however, is disillusioned with his road, 'the great raw cut extending through the forest as far as the eye can reach, until on the horizon it becomes a mere nick in the dark sky-line, like the back sight of a rifle.'[2]

When the road is established, progress brings to Fada confusion, revolution and crime. The old ways begin to go, new conflicts come with new opportunities. The makers of the road have made trouble for themselves. Rudbeck is annoyed to find himself the blind instrument of changes which he does not approve. And his clerk, Mister Johnson, falls victim to the altered times, murdering the white storekeeper for cash and gin, and being shot by Rudbeck for his crime. By bringing in the temptations of civilisation, the line that the road draws through the forest destroys those who were once protected by the haphazard bush in their undemanding ways.

That interesting explorer at the boundaries of cybernetics, Gregory Bateson, has pointed out the problem in trying to 'improve' the savage as well as in trying to leave him alone as an ethnic oddity or anthropological experiment. 'There are two forms of colonial administration,' Bateson declared. 'There is the form of colonial administration which says that the natives have got to be like the colonists. This is missionary endeavour, all that, and becomes a tyranny. The other form of colonial administration says that the natives have got to be like themselves and had better not change. "They have such a beautiful sense of rhythm." Then poetry freezes and everything dies and the flowers can't make seed and nothing goes. So neither of these will do. To do either becomes imperialism.'[3]

Bateson discovered no solution to the problem of spoiling through education and coercion, or condemning to ignorance through refusing to intervene. He merely demanded the right choosing of influences and interactions in the continuing process of man's development in harmony with nature. If this were so easily selected, the end of the imperial regimes would have been less bloody and paradoxical. In fact, the new national governments of Asia and Africa seemed as heaven-sent and hell-bent on eradicating the savage as any European missionary in the bush. The new state of Vietnam was determined to eradicate the communities of the *montagnards*, not only because they had often acted as mercenaries for the Americans, but because their primitive ways were an offence to Marxist thinking – Marx, after all, did condemn the idiocy of rural life. In the African nation of Zaire, President Mobutu stopped anthropologists and tourists from visiting the pygmies of the equatorial forest. He declared that the pygmies did not exist any more, since they demeaned the image of the modern African state. One of the primary policies of most of the elites that governed the emergent nations of the Third World was to present themselves as modern and to ignore their special and primitive heritage outside the field of art, where Western appreciation had made it respectable. Only the Khmer Rouge, however, in a parody of Marxism, sent the urban population of Cambodia to the Killing Fields in a genocidal frenzy, a regression into barbarism, a puritan attempt to put back the clock to a time before the corruption of commercial civilisation. It achieved nothing except pyramids of skulls, more than a million dead, dark savagery with automatic weapons.

Although economic imperialism still has decades to run, political imperialism is ending, even in Eastern Europe and the rest of the Russian Empire. Yet the divisions between the rich and the poor, the developed and the underdeveloped countries and peoples are sharpened, if anything, while the division between the civilised and the savage is curiously reversed. It is the industrial nations of the northern hemisphere which have spread the gospel of conservation and the need for ecology. They can afford to spread this belief as they have passed through the primary stages of industrial development; they are now concerned with preserving the small areas of their own surviving wilderness for urban leisure; they also wish to save the large remaining open spaces in other countries for tourism and for future natural resources. Yet considerations of pollution, of the ecological balance of nature, and of the conservation of primitive ethnic groups are largely irrelevant to underdeveloped countries, which still cannot feed or clothe or house their populations adequately. The paradox of the old imperial powers is that, now the famines and shortages of the Third World are no longer their direct political concern, they can afford to praise and pay for the preservation of those marginal cultures, which the new nations of the Third World will eradicate or include as surely as the Romans did the Iceni, the Americans did the Indians, or the Boers did the Hottentots.

Where there is a deliberate modern policy of the segregation of tribal life and a political doctrine of the separate cultures of black and white, it is universally condemned. South Africa, with its doctrine of apartheid, could cynically claim to be in the forefront of conservation and ecology; its National Game Parks were, indeed, the envy of the civilised world. Its policy of separate homelands for the Bantu peoples could be said to be an effort to preserve their distinct cultures. Of course, such a policy was based historically on the doctrine of a superior Boer race and an inferior native race. In point of fact, the savagery of the Boer massacre of the Hottentots and the Matabele had been matched by the savagery of the Zulus under Tchaka in slaughtering at least a million of the surrounding black tribes. The only question for white apologists was how to excuse the one and condemn the other. The best way of doing so was to claim that the white action was meant to protect the weaker black tribes from the stronger ones and from self-inflicted wounds. As one High Commissioner of Bulawayo wrote in 1932, the Matabele hegemony south of the Zambesi had been a veritable curse to all around them, thus 'on humanitarian grounds alone their complete extermination had become a crying need.'

After partial extermination, there must be assimilation or segregation of the survivors. The philosophy of assimilation presumes a minority which time will include in the conquering majority. Yet where there is minority rule, as in the case of the Boer government of South Africa, assimilation was impossible as a policy and segregation the best method of keeping power. Thus an old-fashioned philosophy and anthropology was invoked by the Boers to justify the separation of the immigrant and the indigenous peoples. 'The important thing,' as Jan H. Hofmeyr wrote fifty years ago, 'is not the native's inferiority, or his equality, or his superiority; what is important is just the fact that he is different from the white man. The recognition of this difference should be the starting-point in South Africa's native policy.'[4] So it became and so it still hangs on, leaving South Africa the pariah of modern industrial nations for putting ideology above economics, and for refusing to admit the essential similarity of men's minds, given half a chance. Although the present government of the National Party has vowed to end apartheid and progress towards black majority rule, a minority of the minority white population will resist all change from the doctrine of supremacy, while the black peoples are still divided among themselves, even in the face of the prospect of power.

Bateson's dichotomy remains. Inclusion and 'improvement' are one form of imperialism, exclusion and apartheid another. The fusion lies in the cautious recognition that a savage and a primitive streak persists in the most civilised of nations and men, and that a particular and isolated culture may still select certain of the discoveries of civilisation without destroying its special heritage. There are possible points of contact and recognition between the literate and the precivilised society despite their natural suspicion of one another; there may be understandings between them without the general aggression and incomprehension that has always characterised their encounters on Turner's frontier between cleared land and wilderness.

* * *

The dialogue about the nature of men in relation to civilisation and the wilderness, which was first recorded in the *Epic of Gilgamesh*, debated by Aristotle, contended by Sepúlveda and Las Casas, disputed by Hobbes and Rousseau, made biological by Darwin and economic by Marx, reached its culmination in the Western world at the beginning of the twentieth century in the beliefs held by the dominant Social Darwinists, socialists and eugenicists. While disagreeing violently on the conclusions to be drawn from their theorising, most of these radicals agreed that the bettering of the human species lay primarily in biological improvement through the pairing of healthy parents, the limitation of the breeding of the unfit, and the reform of the rearing of children. As a matter of intellectual concern, eugenics was the ecology of the early years of this century.

Hitler and his doctrine of the Aryan race and his genetic experiments on human beings caused a revulsion among liberal thinkers. Environmental and cultural factors were now assigned to an overwhelming importance in the education of the human infant, while inherited and genetic characteristics were downgraded. Behaviourism reigned, biology bowed. It was rare in the twenty years after the Second World War to find any leading thinker who held that the nature of mankind was largely determined by innate patterns inherited from the primates or the primitive tribes. Cultural tinkering and piecemeal reform were the fashions and experiments of the day. Society seemed capable of quick advance through applied methods. There might be a speedy global advance and solution to poverty, if the men at the top put their minds to it.

The counterattack of the biologists and the monist thinkers became popular in the early 1970s with the growing realisation that all the laws and reforms of the post-war decades had only increased the gap between the rich and the poor nations, between the educated and the illiterate, between the urban world and those struggling for survival in the barren areas of the Third World. Even the vaunted social legislation of the 'new society' in the United States seemed to stall in front of the stubborn wall of poor white resistance to change, of urban ghettoes and the gathering clouds of economic depression. A succession of popular books, comparing innate human characteristics with those of other primates, seemed to show a widespread need to feel forced into the pattern of one's life rather than to feel responsible for changing it.

Konrad Lorenz, who had himself done genetic work under the Third Reich, led the way with his stylish book *On Aggression*. He argued persuasively that men needed outlets for their inherited aggressions, both as descendants of carnivorous apes and in order to satisfy their instincts and sense of ritual. Robert Ardrey in *The Social Contract* and *The Territorial Imperative* and Desmond Morris in *The Naked Ape* followed Lorenz in this method of approach, finding parallels of behaviour between men and primates, while rejecting behaviourism itself, which claimed that any child could be shaped into any mould by the expert use of rewards and punishments, pleasure and pain.

In a sense, the argument between those who put emphasis on inheritance and those who stress environment as the determining factor in the making of man has been continuing since human thought has been recorded. If the civilised Gilgamesh was held to have conquered the wild Enkidu, it was because Babylon had conquered the savage tribes of the plains and the mountains. Yet faced with a superior civilisation which they destroyed, the Mongols still had no doubt that they were born to rule, and that they had inherited the steppe without need for urban schooling. It is the privilege of the Mandarin on the wall to praise the virtues of a long training, while the Mongol at the gate will yell the importance of the cradle and the saddle. Victory and fashion decide the winning argument.

Edward O. Wilson's invention of the method and his popularisation of the term *Sociobiology* provided a serious attempt to find a common ground of disciplines, in which the various social sciences might come to balanced conclusions. His theory was as evolutionary as his subject, and the fashion for it did desert him with the next surge in human hope, the hope that drastic reforms and behavioural manipulation might still change the human species. Wilson himself included many cautions against misusing his methods to prove that much was innate in mankind, little mutable without benefit of millennia. Yet his findings and suggestions were a valuable corrective to any view which held that the savage in human nature was easily controlled or changed, or that the conflict between savage and civilised societies was between two different sorts of men.

Sociobiology did not aim to draw analogies between different types of animals and human behaviour as in the books of Lorenz and his followers, but to devise and test basic theories about the hereditary basis of social behaviour. Interestingly enough, Wilson came out on Kropotkin's side rather than Darwin's; he emphasised throughout his book the altruistic part of man's actions and his sense of play, opposing Lorenz's stress on man's bent for aggression and murder. Even so, altruism only contributed to genetic fitness, the chance of having offspring. Wilson even valued anti-social behaviour, stating that the act of a 'lone wolf' or of a Timon of Athens in separating himself from his group and living solitary in the woods may have led to the progress of mankind, originally in showing to our hunting ancestors the way from the forests to the dry plains, full of game. He pointed out the factors in man's inheritance which favoured co-operation, such as defence against predators and competitors, group methods of finding food and bringing up infants, a fondness for play in company. He saw sociobiology itself as a peacemaker in the quarrel between behaviourists and biologists, between those who vaunted men's sense of aggression and those who supported men's collective unconscious, between those who wished to conserve what was savage and those who sought to control it.

For Wilson ultimately recognised the dangers of sociobiological thinking, as anyone of his profession should after the totalitarian misuse of genetics in the 1930s. He put the problem succinctly: 'I am aware that the very notion of genes controlling behaviour in human beings is scandalous to some scholars. They are quick to project the following political scenario: Genetic determinism will lead to support for a status quo and continued social injustice. Seldom is the equally plausible scenario considered: Environmentalism will lead to support for authoritarian mind control and worse injustice. Both sequences are highly unlikely unless politicians or ideologically committed scientists are allowed to dictate the uses of science. Then anything goes.'[5] And then Huxley's *Brave New World* would come about.

This discussion of a recent attempt to reconcile the misunderstandings between the various disciplines engaged in the proper study of mankind is partially dictated by the need to explain the ordering of this historical inquiry into the word 'savage' as applied to the environment, animals and humanity. As Wilson points out, from the macroscopic point of view of the visiting zoologist from another planet, the humanities and the social sciences shrink to specialised branches of biology.[6] Social history and its method, however discursive, however inclusive of geography and poetry, of fiction and philosophy, or even of the descriptions of sociobiology, can only be a stimulating process which may provide research materials for those studying the behaviour of *homo sapiens* on his planet. That is good to know.

In our common search to understand something of the nature of man, it may be said in defence of the allusive verbal method, as opposed to the sociobiological method of devising tests between primate and human behaviour, that the grouping of an army of words round the search for the meaning of one word does seem to satisfy the most important attribute that separates men from beasts, the use of language. If Chomsky is correct and there are universal principles underlying the structure of language, if Lévi-Strauss is right and language began in mythical poetry and so helped to form primitive society and our own, then the writing of a book, which tries to avoid technical terms in the hunt of one word that has often been used to describe much human misunderstanding of the environment, may be as valid a method of approaching the problem of mankind on earth as sociobiology or psychoanalysis, or even the *Epic of Gilgamesh*. The subject of this inquiry is the change in meaning of a word in the minds of men, who themselves have changed from living in the forests to dwelling in great cities. It is merely another way of seeking to find the connection between the trees which once covered the earth and the nature of the people who once lived with the beasts in them.

An English police officer, serving in Nigeria this century, was brought an Ibo constable, who had run away from his duty because he claimed that his tree was calling him home. The tree was named an Oji tree and his own name was Oji; in the night, he had heard the tree call to him and he had to go. Such behaviour was no longer considered by anthropologists as mad or even abnormal. It might seem illogical or even poetical, for European poets were quite capable of speaking of the call of a tree. To act upon this call might seem odd to some, but it did not mean that the Ibo as a primitive man could not distinguish between the call of a comrade and the call of a tree; he might feel that both calls were important. The irrational call of glory or honour was capable of making some Western men perform actions as odd and anti-social as the Ibo's effort to return to his tree.

The police officer further noticed that the other native constables judged the Ibo's actions according to their own cultures. The Yoruba, who did not think trees animate, thought that the Ibo was a fool, while the Ekoi thought him sane, since there were sacred trees in their land. Thus the Ibo's action was as reasonable as any holy or ritual act, for a totem relationship between a man and a tree was based on real ties in the past, when men depended literally on trees for their lives.[7]

In northern Europe, tree-worship was as common as it was recently among the Ibos. In Christianity itself, the tree is the symbol both of sin and of pain. In the Garden of Eden, the serpent descends the tree to give Eve the apple and the knowledge of good and evil. Once she has eaten the fruit of the tree, she and Adam wear leaves because they are naked and ashamed. Because of the evidence of the leaves, God expels them from Eden into the wilderness to toil for their living. The tree also represents the Cross which crucifies Christ, as well as the spruce tree of Christmas which heralds the birth of the infant Jesus in a wooden manger. By religion and myth, the forest has influenced the perceptions of modern European man in ways that he hardly recognises, even as the equatorial jungle still affects those who live under its shade.

Psychiatric work among West Africans has suggested that, in the case of the preliterate peoples of the forest belt, sound played the role that sight played to a European.[8] In the humid and gloomy labyrinths of the equatorial forest, the cultures of the blacks depended more on the ear than the eye. Drums were necessary for signalling to the unseen friend, where words could serve in more open country, and a sign language in the desert, such as the Indians developed on the great plains of North America. Sound was the warning of the approach of game or the enemy through the thick trees. Sound was the herald of feast and dance and war. The inventiveness which other cultures put into the visual ingenuities of alphabets and pictures were put by the tribes of the equatorial forest into the subtleties of the drum tap and beat, which broadcast a language and a meaning through the darkest places on earth.

'We penetrated deeper and deeper into the heart of darkness,' Joseph Conrad once wrote of the journey of a Victorian steamer down the Congo River.

> It was very quiet there. At night sometimes the roll of drums behind the curtain of trees would run up the river and remain sustained faintly, as if hovering in the air high over our heads, till the first break of day. Whether it meant war, peace, or prayer we could not tell ... We were wanderers on prehistoric earth, on an earth that wore the aspect of an unknown planet. We could have fancied ourselves the first of men taking possession of an accursed inheritance, to be subdued at the cost of profound anguish and of excessive toil. But suddenly, as we struggled round a bend, there would be a glimpse of rush walls, of peaked grass-roofs, a burst of yells, a whirl of black limbs, a mass of hands clapping, of feet stamping, of bodies swaying, of eyes rolling, under the droop of heavy and motionless foliage. The steamer toiled along slowly on the edge of a black and incomprehensible frenzy. The prehistoric man was cursing us, praying to us, welcoming us – who could tell?[9]

The cultures of the rain forest sounded savage to the cultures from the European cities. Yet the ancestors of these colonial Europeans had themselves come from forests, which they had cleared in order to provide themselves with the original materials for their civilisation and their ships and their written languages. The cold and deciduous and slow-growing woods of the north, split up by mountains and occasional barren spaces, seemed to offer more scope for removal and denial than the inclusive and enervating equatorial jungle with its quick creeping growths. Just as the deserts might have encouraged the study of mathematics and astronomy among the ancient Egyptians and the Arabs because of the nightly opportunity for contemplating the clear heavens and conceiving straight lines and geometrics between the stars, so the rain forest had probably helped to foster the drum culture of the slight variation of sounds, while the timber culture of Europe might well have stimulated communications through words, spoken and written, in the frequent clearings and barren outcrops among its spaced trees.

Yet too much can be made of the connection between human environment and human culture. Certainly, human beings have largely changed their surroundings in northern Europe and have adapted themselves fairly well to life within their new farms and suburbs and cities. If there are residual desires and obsessions left over from an ancient life in the timber, it is a life which they are no longer adapted to endure. Yet what is certain is that all life on earth depends even more on the survival of the last of the primal forests. In this last decade, mass consciousness of the value of the breathing trees has grown. The bad word 'savage' has become the good word 'green'.

There have been popular prophets crying in the wilderness, particularly Rachel Carson in *Silent Spring* in 1962, which aroused global concern about the persistent and continuous poisoning of the whole human environment. But there were other voices crying out, which were hardly heard. At a Commonwealth Forestry Conference in 1974, the main speaker advised that the tropical forests would only remain standing for another thirty years, given the prevailing demand for hardwood. He spoke of man's ancient alliance with the forests and of his present reliance upon them to produce not only the timber and pulp for housing and paper-mills, but also the photosynthesis so necessary for the ecology of the earth. He quoted an Indian study, which linked the nature of the forest to the coming of the monsoon, and he suggested that men would alter their weather patterns as well as their breathing, if they destroyed the last of the great woodlands. He told the assembled foresters that they were the primary guardians of continued life on this planet, for man still depended on woodlands far more than he now recognised, living apart from them. The foresters should carry with them 'the pride of the tree which cannot speak for itself.'[10]

Such a speech, in language not too different from the understanding of the Ibo for the forest, was praiseworthy; but it did not affect the assault on the last great equatorial forest outside the Congo, the Brazilian forest of the Amazon basin. There the government was pursuing the same policy of genocide and the quick waste of natural resources that every civilised government had pursued in history, with the sole possible exception of the British administration of Canada. This was not the first attempt to exploit the Amazon for the world market. There had been the extraordinary rubber boom of the nineteenth century, in which an opera house had been built in the interior city of Manaos and nearly a million immigrant *seringueiros* from the arid *sertão* had measured out their solitary, poor, nasty, brutish and short jungle lives in the balls of latex bled from the wild rubber trees. Sir Roger Casement had been sent there after his report on the Congo to deliver the same verdict on this other bloodthirsty exploitation of men for the sake of tapping profits from the trees; the seeds of these had already been stolen by the British and transplanted in neat rows in Malaya, creating enough competition to break the Brazilian rubber trade even more certainly than the grim verdict of Casement's second report.

The modern assault of the Brazilian government on the Amazon basin began with President Kubitschek's successful folly, the creation of Brasília. If Philip the Second of Spain had decreed the building of Madrid in the centre of Spain to unify its disparate parts, so Kubitschek could decree the building of Brasília in the heart of its territory, to stop its people clinging to its sea-coasts and to make them exploit the interior. The construction of the new capital in the hinterland did provoke the incursion of billions of dollars of foreign investment to develop the wilderness. The jungle to the north was parcelled out into areas, sometimes totalling millions of acres for one company or one man. The great Transamazon Highway was begun with free smallholdings granted along its length to attract settlers and to relieve the pressure of overpopulation in the coastal slums and on the dry *sertão*.

Land clearance came first. The Brazilian pioneers repeated the same sad story of the liquidation of the primitive Indian tribes that had been practised since the first arrival of the Portuguese five centuries before. Epidemics were begun inadvertently or by gifts of infected clothing or food; Indian villages were dynamited or machine-gunned from the air; alcohol and forged documents were used as the instruments of dispossession; forced slavery to the land and the prostitution of the women were the lot of those tribes which did not flee deeper into the forest, but surrendered. The very organisation set up to protect the aborigines, the Service for the Protection of the Indians, became the chief instrument of their extermination, until another government report confirmed this fact and led to a reorganisation of the department, doomed to repeat its process of enforcing civilisation by one method or another. Only on a few reservations, such as the area controlled by the Villa Boas brothers on the Xingu, were the Indians safe, for the time being.

Of the two million Indians who were living in Brazil when the white men first arrived, perhaps two hundred thousand now survive. This latest destruction of savage tribes appears more shocking now, as there are few preliterate peoples left to engage the concern of the wealthier nations, which have relegated the guilt of their own past dealings with savages to their historical conscience. The extensive reporting of this fresh injustice by civilised people towards aborigines has not stopped the process, merely publicised it. It has made small difference that Lévi-Strauss himself protested in a letter, stating: 'Unless we respect man in all his humblest forms, in all his beliefs and customs, however offensive or strange they appear to us, it is humanity itself which we dishonour, and which we expose to the gravest perils.'[11] Just so Las Casas once tried to save the Indians of Mexico and Peru from their exploitation by the colonists, just so he was hardly heard.

The great explorer Humboldt himself had been a child of the Enlightenment and a believer in the equality of men. He had been enraged at the harsh treatment of the Indians by the Christian missions in the jungle, which had used the converted natives as slave labour. 'To say that the savage, like the child, can be governed only by force,' he had written, 'is merely to establish false analogies. The Indians of the Orinoco are not great children: they are as little such as the poor labourers in the east of Europe whom the barbarism of our feudal institutions has held in the rudest state.'[12] He would no more have attempted to kill off the last of the Indian tribes and to destroy the life of the forest than he approved of the little attempts of the Christian missions to establish themselves in the wilderness.

Yet the missions failed, as the latest attempts to ranch and till the poor soil beneath the equatorial forests will fail. Doubtless in the end, with more and better machines, with larger and more ruthless capital backing, the Amazon rain forest could fall, and the Congo too, for the exploitation of their timber resources. The last of the Indian tribes could be preserved as curiosities in a terminal reservation on the Xingu or elsewhere, or they could be assimilated into the melting-pot of the races that has already made up Brazil. For the huge extent of miscegenation on the colonial plantations has always modified the enormous social distance between the Big House and the tropical forest and the slave hut or servants' quarters.[13]

The system of the Big House with its surrounding dependants was, indeed, the only successful civilised area in the tropical forest. As Spengler has written, a people does not migrate from one continent to another, for it cannot take its environment with it. The Portuguese in Brazil set up a hybrid culture adapted to their new environment; but that system was more African than European, based as it was upon imported and home-bred slaves as a method of labour, upon the plantation as a method of agriculture, and upon polygamous patriarchism as a method of family life. On to these features the Portuguese grafted the Catholic religion and a political system of group politics called *compadrismo*. This hybrid culture did endure in Brazil. 'Yet it was Europe reigning without governing: it was Africa that governed.'[14]

Legal slavery has passed, economic slavery abides. Exploitation of the forest and its peoples endures. If only a modified version of the Big House system — great plantations or collective farms, strung out along the rivers or the jungle tracks, living in and off the environment — may work as a civilising force in the equatorial jungles, then perhaps the last forests may not fall before the machines of man. Their survival is possible; but it is only probable if there is an economic collapse in the Western world, a nuclear war or a widespread refusal to leave the megalopolis for the hardships of clearing the land. Even without catastrophe, global ecologists may win the guardianship of these vital harmonisers of the atmosphere of this planet, while nature herself may have the last word against the attempted devastation of man. Even the defoliation of the jungles of Vietnam by Agent Orange will be covered by the forest again within twenty years.

For finally, how little have we altered or scratched at the surface of the earth in all our efforts since we first settled in cities. How quickly the spoiled ground is cloaked with the trees and bushes of nature, once we have turned our backs upon it. How few are the millennia since man has called himself civilised, compared with the millions of years of his slow evolution from an original state.

The urban savage entered the heart of the European and American city in the industrial slums of the last hundred years; but he has now surrounded the exploding cities of the Third World. The *favelas* of São Paolo, the *bustees* of Calcutta, the *villas miserias* of Buenos Aires, the *barong-barongs* of Manila have crept over the rubbish-tips and railway embankments near the cities, until tens of millions of immigrants from the countryside now live in the clustering shanty-towns outside the main urban areas, unwanted, unregulated, dangerous. The teepee of the wandering Apache, the sod hut of the pioneer on the prairie were no worse housing than these shacks of mud wall and corrugated iron roof, of discarded planks and junk, that huddle cheek by jowl outside the new metropolis. Without water and without drains, ruled by gangs and liable to levelling and eviction by the police and the army, these hordes of squatters look for an unlikely opportunity in the overloaded cities that are breaking down from their lack of money and services.

Studies of rabbits suggest that aggression in man is increased by overcrowding.[15] The closer we are packed, the more likely we are to show our teeth and claws. Calcutta is now a slum sixty miles long by the Hooghly river; its density of population was the highest recorded in the world in 1921, and has increased to four times that figure. Over ten million people live largely without sewers and without water; one-tenth of them sleep in the streets; some half of them live below subsistence level. Their numbers are expected to grow to twelve millions by the end of this century. The city administration has no prospect of improving conditions, only of trying to prevent them from becoming worse in the central areas. Although the urban mob has been strangely quiet thus far, mass violence is expected with the next flood or famine or epidemic. Calcutta has bred a likely catastrophe into itself.[16]

Lima is in much the same condition. A majority of the population of that city is now squatting in the *barrios* of the surrounding desert. With or without the guerillas of the Shining Path, they are likely to rise in their time and descend on the privileged centre of Lima in the way that Jack London forecast the People of the Abyss would descend on the West End of London. The riches of empire and world wars and social reform may have prevented such an attack in England; but there is no imperial wealth in Peru nor Bengal to pay for social reform, while another world war will merely spell the end of the megalopolis. In these foul new conglomerations, mankind is breeding the new savages, a underclass with nothing to lose, the shock troops of a new anarchy or vengeance.

Overbreeding and the flight to the urban warrens and *bidonvilles* represent the largest immediate threat to the continued life of mankind on earth, for it will still take decades to deplete the ozone layer, to pollute the oceans, and so to strip the soil of trees and vegetation that the survival of all is threatened. We have made our new labyrinths of injustice and aggression from tarpaper and cartons, from chipboard and packing-cases. Once we fought the savage in the woods, now we have cut down the woods to house him at our doors. While most of the surface of the earth is still wilderness, while the seas are hardly cultivated and the crust of the globe scarcely punctured, we persist in crowding together, fleeing from the isolated squalor of rural life to the terrible promiscuity of the new megalopolis.

Savageness begets savageness. The fear which made the solitary pioneer strike at the stranger or the painted Indian to avert a possible attack was an individual act; but the assault of the mob under the influence of the demagogue is a bloodier thing, for it has the lack of responsibility of the wild animal. As Hemingway knew, in bull-fighting, the crowd is the beast. The terrible sociability of men, the instinct of most of them to pack themselves together for working and living seems to loose more violence upon humanity than when they lived in family or village groups, like the last of the Stone Age tribes today in New Guinea, whose wars are continual but playful, and whose casualties are small.[17]

Even the Russian regime has failed to force or to encourage a large emigration to Siberia in order to take the pressure of population off Moscow and Leningrad and other urban areas. The impending break-up of the Union of Soviet Socialist Republics is partly in response to the domination of the great cities and the pollution of the environment by centralised industry – the whole Aral Sea a stinking mess of poisons, lethal chemicals spread on the soil or floating in the air. The metropolis of the world seems doomed to grow into a global megalopolis, where overcrowding will endanger its own desperate response and reversion to savagery. It is indicative that urban governments and secret policemen have taken to the most refined uses of torture in order to combat guerrillas who have left the banditry of the mountains for the kidnappings of the streets. Once the exquisite torments used by forest Indian tribes such as the Hurons were the terror of the towns of New England; now the experts in inflicting pain are specialists in counter-insurgency, sent out across the world under aid programmes to teach the techniques of breaking down human beings into pliable animals.[18] If the Victorians once considered that torture was the abomination that set mad savages such as Theodore, Emperor of Ethiopia, apart from the civilised human race, now torture is at the service of the more advanced nations as the tool of power and reason and stability in the way that once it served the religious fanaticism of the Inquisition or the tests of bravery of the Hurons.

THE NAKED SAVAGE

With the growth of violent techniques in civilised governments, with the rising rate of crime in most cities that are surrounded or infiltrated by an urban jungle of slums, savagery has become almost acceptable to urban man, a part of his existence and character to be recognised, if not always praised. As the personal terror of the scale of government grows, as the feeling of helplessness of the individual before the computer and the state increases, so the admiration for the savage augments. From Hemingway through the Second World War writers to Norman Mailer, the exaltation of the simple barbarian who can slough off civilisation and let his passion run wild has found a wide and vicarious audience. In *An American Dream*, Mailer's hero Rojack gives way to every impulse of lust and aggression before leaving towards the jungles of Yucatan, back to the beast. Scorn of civilisation is now a literary virtue which would have seemed intolerable to any actual pioneer on the frontier, just as the cult of cruelty in modern drama would have seemed devilish to settlers in Indian territory, longing only for peace in the daily insecurity of their lives. As de Tocqueville once noticed, living in the wilds, the pioneer only prized the works of man.

Because of the revolution in technology and modern life, the fear of the savage in Western societies has been replaced largely by the obsession to recapture him. White man's peyote churches are founded to join in tribal communion, self-expressive dancing imitates the tribal ceremonies and release mechanisms of preliterate peoples. The computer in the skyscraper has begun to conjure up the same terror of soulless destruction that the Celts once felt, living on the plains under the daily horror that the sky would break apart one day and fall on their heads. To cater for such fears in democracies, the supermarkets have begun to stock 'organic' foods and 'natural' cosmetics for the suburban masses, afraid of chemical poisoning. The ecological balance has become a factor in the political process, although all the 'green' parties appear to do is to delay or frustrate economic change, not abolish it. The greenhouse effect is deplored, but carbon emissions will take decades even to limit to present levels. Toxic wastes are dumped at sea, nitrates foul the furrows, acid rain withers the forests, dolphins die in the long killing nets with the tuna – and yet prevention lags far behind proliferation. It is the same tragedy with the spread of nuclear and biological weapons. Ecology and the conservation of energy and of the earth are the sops to the savage dream within us, but they are swallowed when economics and strategy dictate that food and heat and goods and defence must come before our ancient nostalgias.

Savageness begets savageness. Long ago, the melting-pot became the solution for the conflicts in American life. The suburb was the compromise between city and country, the naturalised American and intermarriage was the solution between native and immigrant, mass education was the meeting-point between a barbaric and an élite culture. Yet when one side picks up arms, so will the other; when one side commits atrocities, so will the other. The advocates of Black Power, for instance, state that the blacks should arm themselves against all whites because some whites act like savages towards blacks. It is the solution of the Westerners such as Hugh Henry Brackenridge, who used to advocate harsh treatment for 'the animals, vulgarly called Indians'; it is the solution of the men who classify many sheep with a few goats in order to shear the whole herd to their own advantage.[19] It is also a solution which runs counter to the whole course of American history, which, although often bloody and racist, has moved inexorably towards the absorption of the classes and the immigrants in one bourgeois whole, where the savage may be limited, if not suppressed.

American man is now an urban man and he was recently a rural man. It would be strange if the psychological shock of trying to find streets as natural as fields or woods did not provoke savage explosions in the cities. Claude Brown's brilliant examination of Harlem, *Manchild in the Promised Land*, showed just how much of the black ghetto's barbarism came from the sudden transplantation of sharecroppers from shacks to tenements. Robert Kennedy was using more than a politician's rhetoric when he stated before his murder: 'We confront an urban wilderness more formidable and resistant and in some ways more frightening than the wilderness faced by the Pilgrims or the pioneers.'[20] Oscar Lewis was showing more than an anthropologist's concern when he demonstrated how the two cultures of riches and poverty, suburb and slum, privilege and *barrio* in the large cities of the Americas were doomed to misunderstanding, friction and violence in the crooked streets of their encounters.

The savage is dead, long live the savage. The frontier has moved from the wild to the slums. With more than half of the population of the industrial nations now living in cities, with the villages of the underdeveloped world emptying for the doubtful opportunities of the shantytown, fear of attack has shifted to the corridors from the forests, to the elevator shafts from the trees. Half-hero, half-devil, the savage is at last recognised where he has always been, within the gates, within each man, the innate reaction from civilisation. We resist the savage and he is ourselves. Therefore we have tried to make him more respectable.

10

GREEN AND DYING

As THE LAST of the great rain forests were threatened and as the last of the Stone Age tribes was hunted down for clearance or television cameras, the savagery once attributed to the forests was finally recognised as the characteristic of the new and future cities. Mrs Mary Jemison's saga of horror at the hands of the Shawnees was as nothing compared with the behaviour of the Droogs in *A Clockwork Orange*. The secret policemen of the global capitals, in trying to extract confessions, behaved worse than any Indian tribe in the torture rituals of proving the manhood of prisoners. The football fans of the world were so violent at the play of their heroes that they had to be separated from the pitches by the wire mesh normally used to keep back beasts from visitors to zoos. The murders committed by English football fans on the Continent caused their clubs to be excluded like savages from civilised European competition. A stroll by night along the careful avenues of Central Park in New York was more liable to assault than any Pilgrim journey through the Massachusetts woods. 'Wilding' was the street term used to describe mugging and gang rape in the green spaces of Manhattan girdled by skyscrapers.

The woods, indeed, have been allowed to grow back over the stony ground near the cities as recreational areas. In Connecticut, for instance, where the forest was reduced to one-fifth of the area of the state in the nineteenth century, it now covers three-quarters of the land around the industrial cities and suburbs as the marginal farms have once again been abandoned to the wilderness. A national forest is now being replanted over the coal mines of the Midlands of England to bring back the legends of Robin Hood in new glades of oak and ash and thorn. For the woods have become the peaceful refuge of the urban masses in flight from the brutality of life in their vast conglomerations. The labyrinths of the wild trees are now a part of a leisure industry, the primitive a means of entertainment, the far places of the earth an escape from city life. To be marooned abroad is now the goal of tourism.

So the savage has changed its habitat, although not its nature. Wherever men live in groups, savagery lies in their behaviour. Like the great Norse serpent of legend that girdled the earth many times and ate its own tail, so men seem to feed upon the aggression in their own natures and to need their violence towards one another in order to hold together their precarious characters. Risk and confrontation, whether actual or vicarious, are part of the drama demanded of urban life. Dullness seems a worse enemy than the possibility of daily destruction.

This human urge for continual danger was paralleled by the very process of the ancient forests. So de Tocqueville noted in the 1830s in his classic *Democracy in America*. He saw that, in the depths of the trees, 'the ruins of vegetation were heaped upon one another; but there was no labouring hand to remove them, and their decay was not rapid enough to make room for the continual work of reproduction. Climbing plants, grasses, and other herbs forced their way through the mass of dying trees; they crept along their bending trunks, found nourishment in their dusty cavities, and a passage beneath the lifeless bark. This decay gave its assistance to life, and their respective productions were mingled together.' To de Tocqueville, the inhabitants of this arboreal struggle for life shared in the savagery of the woods. 'The Indian was indebted to no one but himself; his virtues, his vices, and his prejudices were his own work; he had grown up in the wild independence of his nature.' If he were ignorant and poor, yet he was equal and free, hospitable in peace, merciless in war. Living in the forest had enabled him to keep the lost classical virtues of Sparta and early Rome. 'The famous republics of antiquity never gave examples of more unshaken courage, more haughty spirit, or more intractable love of independence than were hidden in former times among the wild forests of the New World.'

Yet the Indian virtue and courage had been useless, when faced with the coming of the European firearm and axe and plough. By the nature of civilisation, if not of man, the savage had to be displaced by the farmer. De Tocqueville's justification of the dispossession of the Indian hunter was ingenious. He followed Locke in maintaining that cultivation created a title in land while a hunting ground was a mere wilderness without title. At the time of its discovery by the Europeans, America could justly be called one great desert. 'The Indians occupied without possessing it. It is by agricultural labour that man appropriates the soil, and the early inhabitants of North America lived by the produce of the chase. Their implacable prejudices, their uncontrolled passions, their vices, and still more, perhaps, their savage virtues, consigned them to inevitable destruction. The ruin of these tribes began from the day when Europeans landed on their shores.'[1] The advance of civilisation, and perhaps even the dispensation of a wise Providence, condemned primitive peoples merely to wait until their inevitable dispossession. They could be admired, but not pitied, for the forest and its savage victims must be cleared by the progress of the town and the field, although de Tocqueville did not foresee that the rise of the urban masses would introduce a worse savagery into the packed metropolis.

In the terms of anthropology and history, de Tocqueville was right; tribal societies had to give way to industrial ones. Yet his wild forest reproducing itself on its own decay and his doomed peoples seemed only seventy years later, to writers such as Jack London and Upton Sinclair, to be the condition of the abysses and the jungles of the industrial slums. Within a further seventy years, there were concurrent and opposed popular visions of future savagery. In one version, world wars and the brutalisation of life would reduce every metropolis to a violent ruin, in which the thought police of 1984 or the Droog gangs of *A Clockwork Orange* would rule the streets. Already in these dark prophecies, Calcutta was a prototype of the ungovernable and starving slum of the future. The other vision of the *Brave New World* by Aldous Huxley saw it as a society controlled by eugenics and the pursuit of pleasure through a drug called soma. Better in Huxley's Utopia to be a pig satisfied than a Savage dissatisfied.

The moralist of *Brave New World* was literally called the Savage. At first fascinated by its scale and easy happinesses, he then took an emetic to purge himself of what he called civilisation. In his dialogue with his Grand Inquisitor, Mustapha Mond, he declared that he liked discomfort and inconvenience; he even wanted the right to be unhappy. 'I want God, I want poetry, I want real danger, I want freedom, I want goodness. I want sin.'[2] He even desired disease and starvation and pain, if he could keep his liberty. When he finally retired to the solitude of the Hog's Back in Surrey to become a hermit and to flog the evil out of himself, the descending tourists with their drug-induced orgy-porgies drove him to a guilty suicide.

To Huxley, there seemed no compromise in 1931 between the Utopia promised by technology and the savagery praised by Malinowski in Melanesia. Future man must surrender to the control of mind and body by the state, or he must live in foul and diseased freedom on some reservation in the wilderness. Fifteen years later, however, Huxley wrote a new foreword to his book, in which he stated that he had found a serious defect in his story. His solitary Savage, brought into the genetic and tranquillised organisation of the new society, had only been offered two alternatives, 'an insane life in Utopia, or the life of a primitive in an Indian village'.[3] At the end of the Depression and the Second World War, Huxley felt that the Savage should have been offered a third alternative – the possibility of a sane existence with a group of exiles from the Brave New World, who would live within the borders of the Indian reservation. Their life would be anarchistic and run on the co-operative principles of Kropotkin. Science and technology would be man's servant, not his master; religion would be his goal. In the atomic and totalitarian age to come, only a large-scale movement towards decentralisation and self-help could save the sanity of mankind. Without it, the world governments would condition their peoples by suggestion, drugs, eugenics and social status to love their servitude. The Brave New World was near, indeed, not six centuries away as in the novel.

By 1958, the pace of social change had so accelerated that Huxley felt compelled to write *Brave New World Revisited*. The organised and totalitarian future already seemed to him to have overtaken the world. He still believed that his pleasure-controlled civilisation was more likely to possess the earth than Orwell's society of *1984*, controlled by pain. The explosion of population, the success of medicine in keeping alive the unfit and the unwanted, must lead to totalitarian and manipulative government in an overcrowded world. 'Death control is achieved very easily, birth control is achieved with great difficulty.' Equally easy to achieve was thought control, through drugs and the manipulation of the mind. The need to keep huge masses alive and packed together in many a megalopolis would force mankind into an organisation modelled on the ants. 'Civilisation is, among other things, the process by which primitive packs are transformed into an analogue, rude and mechanical, of the social insects' organic communities.' The last savages among these human heaps of termites would be the governors, for their thoughts would not be wholly determined. The ant-like society might need the wild or innovative mind at its head to change its course in order to survive. 'These upper-caste individuals will be members, still, of a wild species – the trainers and guardians, themselves only slightly conditioned, of a breed of completely domesticated animals.'[4] In fact, the last refuge of savagery would be aristocracy.

So Huxley took the paradox of the savage to its conclusion. In a controlled society, brutish or rebellious or anti-social behaviour could only be permitted to the very few and the very best at the top of the government. If political liberty were the highest gift of civilisation – the ability to play the beast or the romantic or the misanthrope at one's whim – then all the striving of men to advance from their nest in the trees to the test-tube baby was only an effort to enable the privileged to return to that first lost anarchy, when the strongest had been dominant as well. From prehistory to the *Brave New World*, still only the few could be free. And if the few pulled out to the peripheries of the wilderness, they might continue to exist by becoming sorts of savages again. Although Huxley's advice has been taken literally by certain Survival communities in the western states of America, where an *Ecotopia* has been suggested, such a return to modern tribalism and exclusion has been seen as a form of *ecofascism*, the liberty of some at the expense of condemning the rest of mankind to death by pollution.[5]

In spite of a few voices extolling a return to the Savage, European thought and example has provoked every modern nation to seek urban and technological change, to desire the straight lines and ledgers of European industry and accounting. Even the revolutionary Frantz Fanon wished that fate on the nature of his beloved Africa. 'What I should like: great lines, great navigation channels through the desert. Subdue the desert, deny it, assemble Africa, create the continent ... To turn the absurd and the impossible inside out and hurl a continent against the last ramparts of a colonial power.'[6] Force is wished against force, the line against the line. Nature is to be denied in the battle against imperialism.

Yet what of the savage in this final struggle for man's dominion over man and the earth? In the areas which are marginal, the savage may still survive – in fact, he may be needed. If there is a world famine, his farming techniques may be highly prized – at least, he has been long used to living on little and existing on poor land. At a global meeting of anthropologists at Cambridge in December, 1974, it was decided that perhaps two million preliterate people survived in seven hundred different groups. The anthropologists were naturally careful not to suggest the continued isolation and preservation of the preliterate, for fear of being labelled romantics or rich patrons of educational experiments on deprived human beings. Yet although they thought that no savage societies could resist some connections with civilisation, the anthropologists decided to abandon their own ethical relativity; they declared that such ethnic survivals should be encouraged to remain different for as long as they could, to serve as an example to urban men. 'The last remnants of the pre-agricultural world provide us with the only direct information we can hope to obtain of the biology of the hunter-gatherer populations exposed to the multiple hazards of the natural environment.'[7] If these last specimens did not remain behind, who could teach the teachers?

Yet the fact of the existence of some savage societies is no guarantee of their virtue. Two important modern studies examined a forest people, the pygmies of the Congo, and a mountain people, the Ik of Uganda and Kenya. The pygmies were found to be hospitable and generous and sociable. Sufficient food supplies allowed them to practise the niceties of human intercourse. The Ik, however, had been forbidden to live as nomadic hunters and had been settled as farmers on poor land. As a result, they began to behave viciously to each other, starving their old folk and their children to death in the fierce competition for food. The environment almost wholly determined the behaviour of the two primitive African tribes, who did not distinguish themselves from their surroundings. 'Just as the Mbuti Pygmies in their lush tropical rain forest regard the forest as a benevolent deity, so do the Ik, in their rocky mountain stronghold, think of the mountains as being peculiarly and specially theirs. People and mountain belong to each other and are inseparable.' Or they *were* inseparable in the case of the Ik, until they were forced to give up hunting and live on famine relief. They lost whatever love they had for their mountain world and were bored, mistrustful and sceptical. They were, like many modern urban men, alienated from their place.[8]

To live like the savage, then, is to suffer the same human happiness and miseries at a lower standard of living. To admire the savage from the standpoint of a civilised man is to hallucinate that a primitive life is a possible way of return. Yet if Gauguin did not succeed in finding bliss in Tahiti, who shall? What the many centuries of conflict and misunderstanding between the civilised and the savage have done is to force a recognition of some common ground between the two, which Huxley's few may occupy if they are let alone. Both the civilised and the savage need teaching by one another, although apparently the advantages are heavily in favour of urban man. Yet as Rousseau noticed, urban man did not think so, despite his many blessings. 'It must appear at least very extraordinary, that Habit should have more power to maintain in Savages a Relish for their Misery, than in Europeans for their Happiness.'[9]

At least now the misunderstandings are largely known and the alphabet and the axe, the trade goods and the plough, the teacher and the ledger are changing or limiting the relics of the savage with more caution. For as the *Epic of Gilgamesh* clearly stated, the savage could never be wholly destroyed. When King Gilgamesh ignored the advice of the wild man Enkidu and lopped off the head of Humbaba, the giant spirit of the forest, he loosed the evils of savagery upon the wilderness, upon the wild beast, and upon man himself, who could not deny the fact of his obsessions and his diseases and his death. What has been eroded now is the Victorian belief in the superiority of civilisation and the virtues of mechanical progress. What has been increased is the ambivalent self-consciousness of modern man. 'The more fortunate our condition,' John Updike has warned, 'the stronger the lure of negation, of perversity, of refusal ... Thus the devil thrives in proportion, is always ready to enrich the rich man with ruin, the wise man with folly, the beautiful woman with degradation, the kind, average man with debauches of savagery. The world always topples.'[10]

The atomic age brought about a growing consciousness of the dangers of the economic and biological imperialism which was conquering the planet. 'This has killed a beautiful subject,' a leading physicist declared after the bombing of Hiroshima and Nagasaki.[11] Scientific knowledge could no longer be trusted to transform the world and make it a better place. Nuclear fission had the power to destroy mankind utterly.

Concern about the effects of radiation spread in the decades to come. The effects on the survivors of the two bombed Japanese cities became broadcast, particularly through a brilliant film, *Hiroshima, Mon Amour*. In America, the Committee for Nuclear Information was formed; in England, the Campaign for Nuclear Disarmament. So much scientific and popular pressure was put on governments that overground nuclear testing was ended and deaths from leukemia fell. Then the publication of 1962 of *Silent Spring* by Rachel Carson alerted its general public to the poisoning of the planet by pesticides and toxic waste. Mankind was contaminating its own environment and fouling its own nest, the signal for the extinction of the species. The synthetic poisons could 'accumulate in the tissues of plants and animals and even penetrate the germ cells to shatter or alter the very material of heredity upon which the shape of the future depends.' In the management of the earth for the benefit of mankind, chemicals should not be used to derange the environment, but other natural forces and balances. Nature could not and should not be sprayed with artificial substances, but altered by the introduction of other living things that countered insects and parasites. The 'control of nature' was 'a phrase conceived in arrogance, born of the Neanderthal age of biology and philosophy,' when it had been supposed that nature existed for the convenience of man in a Stone Age of Science. The processes of biology could only be assisted, never prevented or denied.[12]

The next alarm call came from those who were worried about the population explosion, which would deplete the remaining resources of the planet and foul it still further. Ecology became the growing science of the final decades of the twentieth century. It seemed almost a rediscovery of the ancient awareness that human beings were integrated with all other living organisms on the earth. A philosophy called holism and a biological theory called Gaia would rise from this consciousness of what savage and primitive mankind had always sensed. The concept of Gaia as expressed by James Lovelock emphasised that the efforts of *homo sapiens* to destroy the planet might only result in its own destruction, for the globe was full of correcting mechanisms, to which individual species were irrelevant. The most important regulators were the prokaryotes, the simplest and the smallest of cells. The Earth was an immense, self-sustaining organism, on which the impact of volcanoes and meteors had already been far greater than any nuclear war would be. This dynamic system had seen various life forms come and go, but it would renew itself whatever was done to it.[13]

This overview did not reassure those who believed passionately that man himself was debasing his natural relationship to the beasts and the plants, which God had given into his care and control in the myth of *Genesis*. The connection between man and other primates and monkeys was stressed. *The Naked Ape* found that man's behaviour was often merely a bold version of a chimpanzee's.[14] And the application of factory methods to farming particularly disgusted the increasing number of people who thought that animals might have some rights as well as be deserving of the pity of humanity. Not only the vivisection of monkeys in a search to trace the sources of human diseases, but also the agricultural veal-houses and pig 'sweat-boxes' and hen batteries seemed to reduce living things to food units reared in prisons, fed artificially and denied any natural liberty. The actual spread of salmonella and 'mad-cow' disease by induced devices within the food chain made the general public see possible danger to itself in the degradation of domestic creatures. Although few would condone the bomb attacks on laboratories and biologists of the fanatics of Animal Liberation, the work of the believers in animal rights did much to change popular opinion. Where once the British people had delighted in the baiting of wild beasts as a form of play, now they joined in the condemnation of baiters of badgers by a modern magistrate. 'All five of you went to the badger set for sport. Having got there your behaviour can only be described as barbaric...

Society today cannot tolerate this sort of behaviour.'[15]

Through pets, the millions of dogs and cats and birds and tame rodents, human beings in cities maintained their relationship with beasts. At least, a general concern if not a blood bond was shown. Each dog should have his home with his 'master': strays were pitied. And most extraordinarily, in Britain, those isles always accused of preferring their pets even to their children, the royal family resumed its ancient role of the interpreters of creation. While the Prince of Wales was known to be passionate about ecology and to speak to growing plants, his father held a life-long concern for wild life, while remaining an advocate of the sport of hunting. He stated that the 'revealed' religions such as Christianity had been wrong to draw people away from the pagan worship of natural phenomena, which had allowed plants and animals to live and die beside humanity within nature's design. This early reverence by the American Indians and Australian Aborigines had been ecological pragmatism. The same 'emotional commitment, a belief that conservation was also the will of God' was needed 'as a religious and spiritual motive to encourage respect for nature in its own right.'[16] The Christian dominion promised to Adam was wrong, the savage appreciation of an organic world must be rediscovered.

The animal paintings drawn in oxides in Neolithic caves had once displayed a human concern for creation to a few primitive people. Now colourful microphotography in wildlife films started to show on television to mass audiences the complexities of the tiniest of organisms in the interaction of myriad systems of being. The children of the world were informed of the mysteries and necessities and beauties of nature. And from that spread of information, there began to grow a greening of consciousness. The word 'savage' was hardly used now to describe the meeting of forests and beasts and men. The word 'green' took its place. And the perception of the struggle for existence which had given 'savage' a bestial name was changed to a view of interdependent chains of life making for the balance and harmony of the planet. The equation that Savage equals War in Nature was altered to Green equals Peace on Earth.

Yet nothing was really changed in nature in the past thirty thousand years except by the hand of civilised man, who had tried to despoil his green planet. As the authors of *Topsoil and Civilisation* pointed out:

> Man, whether civilised or savage, is a child of nature – he is not the master of nature. He must conform his actions to certain natural laws if he is to maintain his dominance over his environment. When he tries to circumvent

the laws of nature, he usually destroys the natural environment that sustains him. And when his environment deteriorates rapidly, his civilisation declines....

How did civilised man despoil this favourable environment? He did it mainly by depleting or destroying the natural resources. He cut down or burned most of the usable timber from forested hillsides and valleys. He overgrazed and denuded the grasslands that fed his livestock. He killed most of the wildlife and much of the fish and other water life. He permitted erosion to rob his farm land of its productive topsoil. He allowed eroded soil to clog the streams and fill his reservoirs, irrigation canals, and harbours with silt. In many cases, he used and wasted most of the easily mined metals or other needed minerals. Then his civilisation declined amidst the despoliation of his own creation or he moved to new land. There have been from ten to thirty different civilisations that have followed this road to ruin.'[17]

He has also poisoned his green environment and has even changed the very atmosphere which has protected him. Nothing has more wonderfully concentrated the mind of civilised man on his destructive capacities than the discovery of holes in the ozone layer over the poles, leading to the 'greenhouse effect'. When the planet warms significantly, the ice caps will melt and flood the low-lying coastal lands of most civilised countries. The chief destroyers of the ozone layer are the burned forests which can breathe no more and the chemical carbons released into the air. To reverse the effect, mankind must go back to the trees and again become 'green' or 'savage'. Seven million square kilometres of new forest must be planted – including a New Forest in the English Midlands – to stabilise the atmosphere, while the remnants of the rain forests must be preserved. Truly, humanity must return to the woods to fill its lungs and live well. As one of the pioneers of the Back to the Land movement, Edward Carpenter, wrote about one solitary beech tree, like the Norse mythological oak Yggdrasil that held up the heavens from the earth, it was 'no longer a separate or separable organism, but a vast being ramifying far into space, sharing and uniting the life of earth and sky, and full of a most amazing activity.'[18]

Such rediscovery and perception was matched in the world of politics. Not only did 'green' pressure groups force all western democratic parties to hammer an environmental plank into their political platforms, but the collapse of Marxist materialism in Eastern Europe and the Russian empire made pollution become a major issue. The Chernobyl nuclear disaster led to demands for secession by the Ukraine. Industrial poisoning was a major incentive in fuelling independence movements in the Baltic Republics. The chemical rape of Eastern Germany was an inducement in the obliteration of its government and amalgamation with Western Germany. Green consciousness acquired political power through the yearning of human beings for a pastoral past. The ancient escape from the savage became the modern urge to rejoin it.

In the Third World of underdeveloped countries, 'green' arguments received just responses. If the rain forest or the rhinoceros should be preserved, the industrial nations should pay for it. Had they preserved their trees or their wild beasts? They had not. They had almost eliminated them. They had industrialised themselves and had become rich. Their insistence that the underdeveloped nations should forgo the same processes and remain poor was mere hypocrisy and fear of competition.

This fair reply was not wholly effective. For larger arguments about human good prevailed from time to time. Even the savage had found how to speak. Through the pictures recorded by television and video as well as the languages they had to learn, the last of the primitive peoples found how to acquire support for their resistance to the progress of 'civilisation' through its own mechanism – publicity. Aborigines used cameras to demonstrate their dispossession. American Indians went to court to recover possession of the lands they had once managed so well. The verbal means by which Prospero had controlled Caliban were employed to evict the white teacher and intruder. Communications proved a double-edged sword.

Millennial preaching of the end of the world had proved an important warning within the Protestant Reformation in Europe. Four hundred years later, books on *The End of Nature* or *Home Country* were equally popular in their prophecies of ecological disaster if there were not immediate conservation.[19] In times of greater faith, Christians had put up with the evils of the Earth in the hope of their deliverance by Heaven through a Second Coming. In the present age of greater belief in human control of the environment, a divine salvation no longer seemed the way out of human mismanagement. The proper study of mankind was now his abuse of his world, and no God would save him from his deliberate crimes against creation.

Despite the growth of 'green' consciousness and political power, there was only a limited hope of returning a materialistic civilisation back to the lesser rewards of living nearer to nature. Small might be beautiful, but large was normal. The majority of the peoples on the planet wanted their cars and their highways, their television and their refrigerators, their processed foods and cheap heating and crowded houses. If they listened to the arguments of the 'green' prophets, they only wanted to hear about the preservation of their recreation. The proof that acid rain from the emissions of factory chimneys was destroying northern forests made governments legislate to control it. The sewage and chemicals which killed life in the sea and contaminated the beaches of the bathers induced administrative measures to limit the dumping of toxic wastes in the oceans. But only when the poisoning of the planet actually affected the views of urban and civilised people was anything done about it. The divorce between the citizen and the savage still endured.

Yet, in the long term, this separation was irrelevant. Of course, in the long term, we are all dead. And in the longer term, human beings will be all dead or evolved or mutated into other forms. What is certain is that the savage, in the long term, will overcome the assaults upon it. 'Green and dying' is a human perception. Green things are still springing, while mankind may conceive itself to be dying. The tiny organisms, which first brought life to this globe, the algae and the lichens, the bacteria and the parasites, these will hardly be affected by the worst devices that civilised man may inflict on the planet for his own comfort and his self-destruction. The savage will proliferate and some beasts will survive, even if *homo sapiens* is stupid enough to commit mass suicide.

At the end of Tennessee Williams's play, *Camino Real*, the road of reality and dream, Don Quixote cried out, 'The violets in the mountains have broken the rocks!' This declaration that green things would break all resistance ended the drama. The Curtain Line had been spoken. It was brought down. The savage will split and bring down all that stands against it. It is the previous word for 'green'. Its history is the story of continuing life on this Earth.

NOTES

THEOLOGIANS ONCE TRIED to count the number of angels who could stand on the head of a pin. This book tries to arrange its sentences round one word, 'savage'. Even so, a book is too short for such a hydra of a word, just as the angels standing on the head of a pin were sometimes thought to be infinite. Yet 'savage' is now a term of importance, convenient for crossing the borders of many attitudes and disciplines. It is a word that confronts the philosopher as well as the biologist, the archaeologist as well as the novelist, the historian as much as the poet. It straddles the ground between a precise anthropological term, meaning preliterate, and a loose adjective of misunderstanding, meaning bestial. It is a condition of life as much as a term of contempt.

Derived from the Latin word for a wood or forest, *silva*, the old spelling of the word was 'salvage' and its meaning merely geographical.[1] But now, in the same way that 'the frontier' in American history may mean variously a place, a process and a style of life, so 'the savage' may mean the wooded land, the relationship between trees and beasts and man, the nature of those who live in the wild, and their fearsome appearance to those who consider themselves to be more civilised. In the same way that 'the frontier' has been useful as a unifying concept to examine a state of settlement and mind, so 'the savage' may serve as a catalyst for looking at the conflict between green things and wild things and human beings, between hunting and farming peoples, between nomadic tribes and urban citizens, and between primitive and industrial cultures before the present time of ecological awareness.[2] Now at last the future of mankind on earth is held to depend on the great chain of animal life and the breathing forests and clean waters. Not always so.

Historically, the English word 'savage' is derived from the French *sauvage*, which had already combined the Latin meaning of forest with the look of some wooded places, gloomy and horrible. As an adjective, 'savage' was then used to describe the nature of the inhabitants of the forest, untamed and ferocious. Thus 'savage' came to be used in the description of human gestures and manners, meaning rude or unpolished, and when the word was attached to the human temper, it meant cruel and aggressive, like a wild beast. The verb 'to savage', which first meant to act like a savage or to make barbarous, ended by meaning to attack and rend with the teeth; for instance, 'the hunting dogs of the Spanish conquistadors savaged the Indian warriors in battle.' Finally, the noun, 'a savage', was compounded from a confusion of all these meanings and came to be a convenient curse in the mouths of settlers, who needed to demean the word as an excuse for their callousness towards the wild life and the people and the cultures which they were displacing. Although anthropologists later tried to use 'savage' in a technical sense, dividing human development into the three stages of 'the savage' and 'the barbarian' and 'the civilised', the word remained an insult on loose tongues.

In this book, 'a savage' has generally signified a living thing in the backwoods or the wilderness, as distinct from a literate man living in a settled community. A savage need not be ferocious any more than much of nature is. Indeed, a savage may be considered innocent or noble in theory and practice; yet ferocity becomes one of his attributes as civilisation advances remorselessly upon him. 'The savage' may also mean a complex of fears about the wilderness and about hunting societies, which may bring out an instinctive aggression even in a man who considers himself civilised. Thus 'the savage' may be an alien habitat where different creatures live; it may be a person, whether brown or white or black, who is ignorant or scornful of civilisation; it may be the attitude to terror provoked by contact with the primitive; and it may be the recognition of the residual cruelty within human nature, however carefully cultivated.

As a whole, this book has tried to distinguish between these meanings as far as possible. For if there is any significance in a term like 'the frontier', through which history may be viewed as a subtle amalgam of geography and philosophy and ideology and anthropology and the analysis of facts, then perhaps an examination of 'savage' may prove a guide to our attitudes towards the primitive and the violent and the wild. Some of this book is based on an earlier work of mine, *The Savage* (London, 1977). The book was prescient on the rise of consciousness about the environment, but it did not have the benefit of modern research into ecology and the balance between plants and wild life and mankind.

THE NAKED SAVAGE

1. It is interesting that T.S. Eliot named one of his Four Quartets 'The Dry Salvages' after a small group of rocks off Cape Ann, Massachusetts, 'presumably *les trois sauvages*'. The poem deals with the permanence of wilderness and the mutability of men and time.

2. As Henry Nash Smith pointed out in his important *Virgin Land: The American West as Symbol and Myth* (Boston, 1950), the creator of the frontier theory of American history, Frederick Jackson Turner, explicitly defined the frontier as the point where agricultural settlement touched the wilderness, 'the meeting point between savagery and civilisation'. I am particularly indebted to *Virgin Land* for my later discussion of Turner's frontier thesis.

   I would like to take this opportunity to express my gratitude to those other historians whose profound and stimulating work on the relationship of geography to history has so much stimulated my own thinking, in particular to Fernand Braudel and Walter Prescott Webb. Furthermore, I must thank those American social historians whose depth and insight and breadth of vision taught me personally as a student that no evidence of social history and no interaction between a human being and his environment is unworthy of consideration. Therefore, my gratitude again to the generosity in time past of Professors Denis Brogan, Oscar Handlin, Richard Hofstadter and David Potter for the many hours spent on overseeing me.

1
WHEN WILD IN WOODS

1. William H. McNeill, *The Human Condition: An Ecological and Historical View* (Princeton, N.J., 1980) is stimulating on the importance of parasites and predators, which he calls macroparasites, in history.

2. Andreas Lommel displays a profound understanding of the mind of the early hunters in his analysis of *Prehistoric and Primitive Man* (Illustrated edn., London, 1966).

3. Sir Edward Burnett Tylor, *Primitive Culture*, was published in 1871 and changed cultural sciences, then in a primitive state themselves. His concepts of the significance of plants and trees and animals, souls and spirits and the dead in early religions and societies has influenced anthropology to this day. Unfortunately, he was Victorian enough to believe that primitive or savage people were part of 'the lower races'. His understanding did not extend beyond that stereotype of his age.

   The bear festivals of the Ainu mentioned by Tylor have been recently studied by the Japanese ethnologist Kindaichi. He stresses that the killing of the bear has nothing to with ritual sacrifice. The bear itself is divine. All animals are gods and inhabit an unseen world rather like our own. They cross over to earth to play. Killing them releases their souls to return to their real home. Their meat and fur are merely the gifts they leave behind for their human helpers.

4. For a stimulating discussion of the upsetting influence of the new archaeology

on the assumptions of the previous archaeologists about the pervasive influence of the Near Eastern city, see Colin Renfrew, *Before Civilisation: The Radiocarbon Revolution and Prehistoric Europe* (London, 1973). He both applauds and attacks the seminal work of V. Gordon Childe, whose *New Light on the Most Ancient East* (London, 1934) remains a provoking and brilliant book. For the ecological approach to our early history, Graham Clark's *Prehistoric Europe, the Economic Basis* (London, 1952) can be recommended, also Robert Braidwood's work on the invention of farming in the Near East. The basic shift of new archaeological school is from comparing artifacts to comparing societies; objects seem less important than relationships between different classes of data.

5. For the rise of patronage from early societies see my *The Need to Give: The Patrons and the Arts* (London, 1990).

6. While the excavations at Jericho are claimed as proof of an urban culture well before the cities of Mesopotamia, they seem only to show a society of grouped villages. The lower valley of the Indus may well have supported cultures which might technically be called urban at this period, particularly at Mohenjo-daro, but proof is difficult.

7. For the translation of the *Epic of Gilgamesh*, I have relied on that of N.K. Sanders in the Penguin Classics series. In his turn he has mainly followed the sequence of Heidel and Speiser, who put together Old Babylonian, Hittite and Assyrian material. Sanders has also added important Sumerian fragments in the episode of the Cedar Forest.

8. The quotation comes from C.P. Fitzgerald, *China: A Short Cultural History* (rev. ed., London 1950). I am also indebted to the work of K.A. Wittfogel, whose *Oriental Despotism* (New Haven, 1957) remains provocative and indispensable, to Owen Lattimore for his theory of the frontier applied to Chinese history, to Professors Goodrich and Needham, and to the revisionist work of W. Eberhard, from whose *Conquerors and Rulers: Social Forces in Medieval China* (rev. ed., Leiden, 1965) comes the quotation from Liao-shih and Ying-wei-chih. Eberhard points out that, in the Chou dynasty, barbarians were thought to be within Chinese culture, since the people of the inner states regarded Ch'u and Wu as 'southern barbarians' and Yen as 'northern barbarians'. Provincialism can build its own barriers within the state.

9. Other great walls, such as Hadrian's and Agricola's in Roman Britain, were defensive; their purpose was to keep out the Picts and the Scots; they did not become the bricks of an exclusive mentality.

10. Adolf Erman's *Life in Ancient Egypt* (London, 1894) remains a comprehensive source (see particularly pp. 500–4) I am again grateful to Tylor's work on *Primitive Culture* for his comparison of Egyptian and Hindu animal worship, from which the quotations are taken.

11. Other than the original Greek sources and the staple works of Bury and Zimmern, I have found insights n H. Michell's *The Economics of Ancient Greece* (Cambridge, 1940) and V. Ehrenberg's *From Solon to Socrates* (London, 1968). Also Esther Boserup's *The Conditions of Agricultural Growth* (London, 1965)

points up the stimulus which population pressure gives to intensive farming and the growth of society as in ancient Greece.

12. Eusebius was quoting the mystic cosmogony of the Phoenician Sanchoniathon.

13. See Ossian's *Halieutica* (my translation), quoted by Mary Midgley, *Beast and Man: The Roots of Human Nature* (London, 1979). This is a stimulating and illuminating work by a moral philosopher on the relationships of the species. It is essential reading for all those interested in the subject.

14. This incident is mentioned in Ivan Sanderson, *The Dynasty of Abu* (London, 1960), also by Pliny in his *Natural History* as well as by Cicero.

15. I am indebted to E.S. Turner's admirable examination of how the British people were shamed into sparing 'brute' creation, *All Heaven in a Rage* (London, 1964), for bringing up Gibbon's comments on the Emperor Commodus. It is also useful on Old Testament concern about treating livestock well.

16. This statement comes from Denys Hay, *Europe: The Emergence of an Idea* (New York ed., 1966), p. 23. This brilliant and modest book traces the growth of one word, 'Europe', and it illuminates ancient and medieval history through its method.

17. See the chapter on 'The Worship of Trees' in Sir James Frazer's *The Golden Bough: A Study in Magic and Religion* (abr. ed, London, 1929), pp. 109–20. Frazer's global learning and lucid style still make him a model for anthropologists.

18. The text of *Genesis* comes from the King James's version, which I have retained. The literal belief that God had given the earth and all living things to men to subdue and dominate was the justification for the spread of Christian imperialism until the modern age.

19. In his penetrating *Man's Responsibility for Nature* (London, 1971), John Passmore quotes and explains St Augustine's interpretations.

20. Aristotle showed a crude and chauvinistic geopolitical determinism in one passage in the *Politics* (tr. B. Jowett): 'Those who live in a cold climate and in Europe are full of spirit, but wanting in intelligence and skill; and therefore they keep their freedom, but have no political organisation, and are incapable of ruling over others. Whereas the natives of Asia are intelligent and inventive, but they are wanting in spirit, and therefore they are always in a state of subjection and slavery. But the Hellenic race, which is situated between them, is likewise intermediate in character, being high-spirited and also intelligent.'

21. All modern historians are in the debt of Fernand Braudel, whose *The Mediterranean and the Mediterranean World in the Age of Philip II* (rev. ed., 2 vols., London, 1972–3) not only gives a complete analysis of the interaction between geography and economics and politics in an important time and place in history, but also changes the very nature of historiography. (For the quotations, see I, p. 187, and II, p. 770.)

22. Even William of Malmesbury's version of Urban II's speech at Clermont,

NOTES

which launched the First Crusade, showed that Christian pride was tempered by a certain feeling of inferiority towards the Arabs. 'They have made Asia, which is a third of the world, their homeland – an area justly reckoned by our fathers as equal to the other two parts both for size and importance ... They have also forcibly held Africa, the second portion of the world, for over two hundred years ... There remains Europe, the third continent. How small is the part of it inhabited by us Christians! For none would term Christian those barbarous peoples who live in distant islands on the frozen sea, for they live in the manner of brutes. And even this fragment of our world is attacked by the Turks and Saracens.' Quoted in D. Hay, *Europe*, p. 32.

23. This passage from Richard of Devizes, *Chronicles of the Crusades*, is quoted by Alfred W. Crosby in his provocative *Ecological Imperialism: The Biological Expansion of Europe, 900–1900* (Cambridge, 1986).

24. For the quotation from Ibn Khaldun, I have used the Baron de Slane's translation of his *History of the Berbers* (Algiers, 1852).

25. Even the Chinese seem to have sent expeditions as far as Madagascar to satisfy the Arab appetite for slaves. A Chinese document of 1178 AD refers to 'an island in the sea on which there are many savages. Their bodies are black and they have frizzled hair. They are enticed by food and then captured [and then sold] as slaves to Arab countries, where they fetch a very high price.'

26. In *The White Goddess* (new ed., London, 1961) Robert Graves has particularly tried to show the existence of a Druid alphabet. He has claimed that in the ancient Welsh riddling poem, 'The Battle of the Trees' (in which the various trees of Britain fight each other for supremacy), the name of each tree actually represents a letter in the ancient Irish alphabet, the birch for B, the oak for D, the yew for I, and so on. Furthermore, he claimed that, in all Celtic languages the word for 'tree' also signified the word for 'letter', because of the Druid use of oak-groves and twigs in their messages. The very word 'book' was connected with the beech tree, since early writing-tablets were often made of beech wood. Thus, for Graves, the worship of the trees became the means of recording that worship; ritual became object.

If writing is, indeed, the prime distinction between savage and primitive peoples, then the progress from tree-rituals to the use of wooden tablets was a giant step towards civilisation. The next giant step from the wooden writing-block was to be the Chinese invention of paper and of printing, developed in Europe by Gutenberg's wooden type, a process which ended in the use of woodpulp for newspaper presses – in Auden's phrase, a matter of 'turning forests into lies'. In that most literal way, the medium was to become the message, the wood was to become the word.

27. I have used the New Temple edition of *The Tempest* by William Shakespeare. The context of the quotation is the first meeting between Prospero and Caliban, in which the subtleties of colonial control are explored.

28. In Paul Radin's searching *Primitive Man as Philosopher* (new ed., New York, 1957), I have chiefly relied on his chapter of 'Freedom of Thought', from which

this quotation is taken as well as the Tlingit poem. But, of course, his book and this book have to be written in words.

29. Noam Chomsky's *Syntax Structure* (The Hague, 1957) remains basic to his thought. Also important to the understanding of his early work is his attack on behaviourism contained in his review of B.F Skinner's *Verbal Behaviour in Language*, XXXV (1959), pp. 26–58. As for Claude Lévi-Strauss, his *Structural Anthropology* (New York, 1963), *The Savage Mind* (Chicago, 1966) and *Totemism* (Boston, 1962) are the more accessible indications of his thinking. The quotation comes from *Tristes Tropiques* (Paris, 1955), p. 421. Benjamin Whorf's recent analysis of the speech of the Hopi Indians in Arizona seems to prove that language tyrannises even the speakers of primitive languages by forcing them to think in habitual patterns. The Californian Yokut Indians, for instance, are incapable of expressing any metaphor in speech.

30. This same complacent attitude has also been attributed to a contemporary of Confucius, Jung Ch'i-ch'i, to Plato and to Socrates.

2
THE AMERICAN SAVAGE

1. The quotation is taken from Gomes Eannes de Azurara, *The Chronicle of the King Dom João I* as reprinted in V. de Castro e Almeida ed., *Conquests and Discoveries of Henry the Navigator* (London, 1936), pp. 81–2. The original manuscript of this contemporary chronicle is in the Bibliothèque Nationale in Paris.

2. Useful for an understanding of the Valladolid at this time is B. Bennassar, *Valladolid au Siècle d'Or* (Paris, 1967).

3. As Braudel is to the Mediterranean in the sixteenth century, so Johan Huizinga is to the end of the medieval era. His work *The Waning of the Middle Ages* (London, 1924) is a synthesis of art, literary and social history which presents an illuminating description of the mood of the time. (The quotations are from pages 10 and 25 of the recent Pelican edition.)

4. This quotation comes from John of Holywood's *Sphaera Mundi* in the 1498 edition.

5. J. Huizinga, *Homo Ludens* (Boston, 1955) remains essential on the subject of play as part of culture.

6. The quotations come from Sir Thomas More's *Utopia* of 1515, and from Montaigne's *Apology for Raimond Sebond* and his *Essay on Cruelty*.

7. There is an approving modern biography called *Bartolomé de las Casas* by M.G. Fernandez (2 vols., Seville, 1953–60), and a disapproving biography by R.M. Pidal, *El padre las Casas: su doble personalidad* (Madrid, 1963). The continual conflict about Las Casas reflects the insoluble contradictions in the nature of imperialism. Las Casas's own *Brevissima relación* has often been reprinted and provides material against the Spanish control of the New World. The best

NOTES

edition of Las Casas's *Historia de las Indias* is by A.M. Carlo and Lewis Hanke (3 vols., Mexico City, 1951). Fundamental for the dispute between Las Casas and Sepúlveda are Lewis Hanke's own writings, particularly his *Aristotle and the American Indians* (London, 1959). Also important is his work *The Spanish Struggle for Justice in the Conquest of America* (Philadelphia, 1949). The most judicious introduction to that situation remains Charles Gibson, *Spain in America* (New York, 1966).

8. F. Braudel, *The Mediterranean*, II, p. 754. Even Bernard Lewis in his understanding *Race and Slavery in the Middle East: An Historical Enquiry* (Oxford, 1990) cannot excuse the role of the Arabs as traders in slaves.

9. Quoted by L. Hanke, *Aristotle and the American Indians*, p. 49.

10. Ibid, p. 47.

11. Ibid, p. 65.

12. See L. Hanke, 'The *Requerimiento* and Its Interpreters', *Revista de historia de America*, I (1938), pp. 25–34. He also quotes the reply of the Bishop of Avila to Queen Isabella.

13. This passage is contained in the letter that Verrazzano wrote to Francis I, 8 July 1524, reprinted in D. Quinn ed., *North American Discovery Circa 1000–1612* (New York, 1971), p. 67.

14. From 'The Voyage Made by Master John Hawkins ... Begun in An. Dom. 1564.' This was written by John Sparke and reprinted in *The Hawkins' Voyages*, The Hakluyt Society, No. 57 (London, 1878), pp. 8–64.

15. See René de Laudonnière, *A Notable History Containing Four Voyages Made by Certain French Captains unto Florida* (tr. Richard Hakluyt, London, 1587) and reprinted in D. Quinn ed., *North American Discovery*, pp. 160–2.

16. From the Temple edition of *The Life of Timon of Athens*, Act IV, Sc. i, p. 71.

17. I have used the Penguin Classics edition of *The Vinland Sagas*, translated by Magnus Magnusson and Hermann Palsson. For a full examination of the early evidence on the Norse expeditions to Vinland, see G.M. Gathorne-Hardy, *The Norse Discoveries of America: The Wineland Sagas*, (Oxford, 1921). In describing the characteristics of Eskimos and Red Indians in the various sagas, the Icelandic poets clearly consider the racial differences between the Skraelings too unimportant to dignify the various tribes with different names. Ari Thorgilsson, Iceland's first vernacular historian and a reliable source, makes an anthropological assumption out of the prejudice of the Norsemen; for he says in his history of the Icelandic people, the *Islendingabók*, written about 1127, that the first colonisers of Greenland under Eirik the Red 'found there human habitations, both in the Eastern and Western parts of the country, and fragments of skin-boats and stone implements; from which it can be concluded that the people who had passed through were of the same kind as those who have settled in Vinland and whom the Greenlanders call Skraelings'. In fact, the first Vikings in Greenland about 985 found the remnants of Eskimo culture there, which they confused with Indian culture; the Eskimos had retreated north into

Greenland with the receding icecap and were to advance to destroy the Greenland colonies as the icecap advanced again five centuries later.

In general, the Vinland Sagas seem to prove that the Norsemen must have reached North America. Although archaeologists have as yet failed to locate remnants of Norse settlements in the United States except for the disputed Newport Tower, Dr. Helge Ingstad has excavated a Norse settlement at L'Anse aux Meadows in Newfoundland. There is too much detail in the accounts of the Skraelings, however, to suppose that the sagas are not based on actual experience. For instance, *Eirik's Saga* describes the rations of five isolated Skraelings discovered by the Norsemen as 'containers full of deer-marrow mixed with blood', the pemmican of the Indian on a hunting trip.

18. On the subject of the Canadian Indians, Jacques Cartier wrote: 'They have no other dwelling but their boats, which they turn upside down, and under them they lay themselves all along upon the bare ground.' (Hakluyt translation).

19. This line from Columbus's famous letter is quoted in an important book on the conflict of cultures at the meeting of Europeans and early Americans in Howard Mumford Jones, *O Strange New World* (London, 1965). The book is one of the few to deal with the Utopian and infernal versions of the Americas which stemmed from the beliefs and prejudices of medieval Europe.

20. I have used an early English version of Antonio de Herrera's book translated by Captain John Stevens and published in London in six illustrated volumes in 1740 under the title of *The General History of the Vast Continent and Islands of America, Commonly called, The West Indies, from the First Discovery thereof: with the best Accounts the People could give of their Antiquities.* (As in other early translations, I have modernised the spelling and the overuse of capital letters.)

21. From the Penguin Classics translation of *The Lusiads* by William Atkinson, Canto 2, pp. 70–71.

22. Amerigo Vespucci operated mainly in Spain and managed to name the new continents after himself, when the Columbias might have been a juster name. This letter of his is translated by G.T. Northup and was printed in Princeton in 1916.

23. Herrera, *General History*, I, p. 277.

24. All these accounts of the reaction of the Mexican Indians to the coming of the Spaniards can be found in Miguel Leon-Portilla's admirable compilation of the Indian chroniclers, *The Broken Spears* (Boston, 1962).

25. Herrera, *General History*, I, p. 339.

26. Victor W. von Hagen has edited a work on Inca Peru taken from the writings of Pedro de Cieza de Leon, *The Incas* (Norman, Oklahoma, 1959). This quotation comes from his judicious introduction, p. lix.

27. Portilla, *The Broken Spears*, p. 41.

28. Bernal Diaz del Castillo, *The Discovery and Conquest of Mexico, 1517–1521* (New York, 1956), pp. 218–19.

29. Herrera, *General History*, II, p. 361.

30. Francisco López de Gómara, *Cortes, The Life of the Conqueror by His Secretary* (Berkeley, 1966), pp. 165, 33. The urbane Edward Bancroft, in his *Essay on the Natural History of Guiana, in South America* (London, 1769), p. 260, probably has the last word on the dispute with the savagery of European massacre compared with Indian cannibalism. 'It is certainly more unnatural to kill each other by unnecessary wars, than to eat the bodies of those we have killed: the crime consists in killing, not in eating, as the worm and vulture testify, that human flesh is by no means sacred. But tho' civilized nations abhor eating, they are familiarised to the custom of killing each other, which they practise with less remorse than the savages. But custom is able to reconcile the mind to the most unnatural objects.'

31. Portilla, *The Broken Spears*, pp. 51, 68. Before Columbus even reached America, Azurara noted the disdain which the Neolithic Canary Islanders had for precious metals. 'They have no gold or silver, nor any money, nor jewels, nor anything else artificial, save that which they make with the stones that serve them as knives, and so they build the houses in which they live.

    'They disdain gold, silver, and all other metals, making mock of those that desire them; and in general there are none of them whose opinion is different. No quality of cloth pleases them, and they make mock of those who desire it ... But they greatly value iron, which they work with their stones, and of this they make hooks for fishing.'

    From Azurara, *The Chronicle of the Discovery of Guinea* (1448), as reprinted in V. de Castro e Almeida ed., *Conquests and Discoveries of Henry the Navigator*, op. cit., p. 227.

32. William H. Prescott, *History of the Conquest of Mexico and History of the Conquest of Peru* (Modern Library ed., 1 vol., New York), p. 967.

3
THE SAVAGE WITHIN

1. Thomas Harriot, *A Brief and True Report of the New-Found Land of Virginia* (London, 1588), as reprinted in D. Quinn, *The Roanoke Voyages, 1584–1590* (London, 1952), I, pp. 317–87. A facsimile edition of Harriot's book with Theodor de Bry's 1590 engravings, taken from John White's original watercolours, was issued by the Edwards Brothers at Ann Arbor, Michigan, in 1931. The watercolours themselves are now in the British Museum.

2. Richard Hakluyt's *Instructions* of 1606 were printed in E. Arber, *Travels and Works of Captain John Smith* (Edinburgh, 1910), I, pp. xxxiii-xxxvii.

3. These quotations come from the observations of George Percy, the brother of the Earl of Northumberland, who briefly served as Deputy Governor of Virginia when Captain John Smith was demoted in 1609. Percy's account was printed by Purchas, *Pilgrims*, IX, ch. 2.

4. See T. Harriot, op.cit.

5. See Verrazzano's letter to Francis I, 8 July, 1524.

6. William Bradford wrote *Of Plymouth Plantation* between the landing of the Pilgrims in 1620 and 1647. It was first published in 1865. Bradford is both a good recorder and unconscious reflector of the Pilgrims' desires and fears. These quotations are taken from the two volumes of the 1912 edition by the Massachusetts Historical Society.

7. See 'God's Controversy with New England', in *Proceedings of the Massachusetts Historical Society*, XII (1871–3), pp. 83–4.

8. Samuel Willard, *A Complete Body of Divinity* (Boston, 1726), p. 604. I have relied for much of my information about the Puritans on Edmund S. Morgan's important *The Puritan Family* (rev. ed., New York, 1966).

9. See Massachusetts Historical Society *Collections*, Fourth Series, VII, p. 25.

10. Rolfe's account is printed in Ralph Hamor, *A True Discourse of the Present Estate of Virginia* (London, 1615) under the title of 'The Copy of the Gentleman's Letter to Sir Thomas Dale that after married Powhatan's daughter, containing the reasons moving him thereunto'.

11. These laws are printed in John Pory, 'A Report of the Manner of Proceeding in the General Assembly Convented at James City ... July 30th–August 4th, 1619', *State Papers Domestic, James I*, I, No. 45.

12. From Bradford, *Of Plymouth Plantation*.

13. William Carlos Williams, *In the American Grain* (new ed., London, 1966), p. 80. While Williams deliberately flouts historical method, his poetic search for the American experience in history has rare insights.

14. Walter P. Webb, *The Great Plains* (New York, 1931), p. 33. This seminal work about the influence of geography upon history has put me deeply in its debt, as has Henry Nash Smith's *Virgin Land: The American West as Symbol and Myth* (Cambridge, Massachusetts, 1950).

15. The words are those of Indian agent B. O'Fallon, quoted by Webb, *The Great Plains*, p. 147.

16. Major Stephen H. Long first called the plains 'the Great American Desert' in his report of 1821. He declared that the area 'is almost wholly unfit for cultivation, and of course uninhabitable by a people depending upon agriculture for their subsistence Although tracts of fertile land considerably extensive are occasionally to be met with, yet the scarcity of wood and water, almost uniformly prevalent, will prove an insuperable obstacle in the way of settling the country.' Although cattle ranching and intensive farming has been developed on the Great Plains, too often the prairie has ended as a dustbowl in the wake of the plough and the wind. And unless the problem of water is finally solved in that arid area, it may yet revert to the Great American Desert of its original name.

17. Captain Randolph D. Marcy, quoted by Webb, *The Great Plains*, p. 158.

18. For the best description of the life of the Mountain Men, see Ray Allen

Billington, *The Far Western Frontier, 1830–1860* (New York, 1956), pp. 41–68. I am much in his debt for his evocative descriptions.

19. The best modern study of the tragedy of the Donner expeditions is George R. Stewart, *Ordeal by Hunger: The Story of the Donner Party* (New York, 1936).

20. I am again indebted to M. Midgley, *Beast and Man*, for bringing the quotation from Plato's *Republic* to my attention.

21. William H. McNeill refers to 'gunpowder empires' in his important *Plagues and Peoples* (New York, 1976).

22. The traveller in the Congo is D. Lopez, who with F. Pigafetta produced *A Report of the Kingdom of Congo* (tr. M. Hutchinson, London, 1881).

4

THE SAVAGE ENSLAVED

For an introduction to the subject of this chapter, James Pope-Hennessy's study of the Atlantic slave traders, 1441–1807, called *Sins of the Fathers* (London, 1967) is illuminating although from a rather personal point of view. More detailed and informative is Douglas Grant's excellent study of Job ben Solomon, the Muslim son of a high priest in Gambia, who was freed by General Oglethorpe and impressed London society on his way home. The book is called *The Fortunate Slave: An Illustration of African Slavery in the Early Eighteenth Century* (Oxford, 1968). Also important are K.G. Davies, *The Royal African Company* (London, 1957) and all of Basil Davidson's works on early African society, especially *Black Mother: Africa: The Years of Trial* (London, 1961), in which the author draws important parallels between feudal Europe and the Africa of the slavers. On the economic side, Eric Williams's *Capitalism and Slavery* (new ed., London 1964) is most important as is the revisionist work by Eugene D. Genovese, *The Political Economy of Slavery: Studies in the Economy and Society of the Slave South* (London, 1966) and *Roll, Jordan, Roll: The World the Slaves Made* (New York, 1975). The controversial work on slavery by Robert W. Fogel and Stanley L. Engerman, *Time on the Cross: The Economics of African Negro Slavery* (Boston, 1974), used a prodigious amount of advanced mathematics and suspect methods to try and show that slavery in the Southern States was an efficient, prosperous and fairly stable system. Later, after many attacks, the authors revised their findings, publishing in 1990 two volumes on their evidence, methods and technical papers, to complement Fogel's *Without Consent or Contract: The Rise and Fall of American Slavery* (New York, 1989).

1. Francis Moore, *Travels into the Inland Parts of Africa with a Particular Account of Job ben Solomon, who was in England in the Year 1733, and Known by the Name of the African* (London, 1738), p. 87.

2. Quoted by D.Grant, *The Fortunate Slave*, p. 128.

3. *The World Displayed: or a Curious Collection of Voyages and Travels* (London 1774–8), introduction by Dr Johnson.

4. John Barbot, *A Description of the Coasts of North and South Guinea* in A. and J. Churchill, *A Collection of Voyages and Travels* (6 vols., London, 1732), V, p. 34.

5. See William Snelgrave, *A New Account of Some Parts of Guinea, and the Slave Trade* (London, 1734).

6. See Mungo Park, *Travels in the Interior Districts of Africa* (London, 1799).

7. The one surviving account of the African trade in flesh by a contemporary African is to be found in D.C. Forde, *Efik Traders of Old Calabar* (London, 1956), in which he prints most of the *Diary of Antera Duke, an Efik Slave-Trading Chief of the Eighteenth Century*, who gave this account of ritual beheading.

8. J. Pope-Hennessy, *Sins of the Fathers*, is at his most eloquent about his visit to modern Old Calabar, while Basil Davidson, *Black Mother*, p. 90, brought the quotation to my notice.

9. W. Snelgrave, *A New Account*, p. 161.

10. Mary B. Chesnut, *A Diary from Dixie* (B. Williams ed., Boston, 1949), pp. 21–2. The diary is both consciously and unconsciously revealing, and the memorial of a wise and witty woman, whose conclusion was, 'Slavery does not make good masters.'

11. Quoted in J. Pope-Hennessy, *Sins of the Fathers*, p. 99.

12. Benjamin Franklin, *Writings* (A.H. Smyth ed., New York, 1905–1907), VI, p. 295.

13. James Houston, *Some New and Accurate Observations of the Coast of Guinea* (London, 1725), pp. 33–4.

14. Le Vaillant, *Travels into the Interior Parts of Africa by Way of the Cape of Good Hope in the Years 1780, 81, 82, 83, 84 and 85* (2 vols., London, 1790), II, pp. 113–26.

15. *Livingstone's Private Journals, 1851–1853* (I. Schapera ed., London, 1963), p. 161.

16. See Sir Samuel Baker, *The Albert N'Yanza: Great Basin of the Nile and Explorations of the Nile Sources* (new ed., 2 vols., London, 1962), *passim*. These two volumes remain the perfect expression of the utility of ignorant self-confidence for survival in savage country.

17. See Sir Richard Burton, *The Lake Regions of Central Africa* (new ed., 2 vols., London, 1961), passim. No biography of this brilliant and complex man has yet managed to explain his contradictions and oscillations between total arrogance and insecurity.

18. I am much indebted to Ronald Robinson and John Gallagher with Alice Denny for their important redefinition of the scramble for Africa in *Africa and the Victorians* (London, 1961), from which the first quotation is taken. They stress the irrelevance of a garrison Africa to successive Victorian governments,

preoccupied with sea power and India and white colonial dominions. The step from influence to imperialism is shown as a result of the persistent crisis in Egypt; the British occupation of the Nile and the Suez Canal began the scramble. The book also points out that the rising nationalism of the Egyptians, the Irish and the Boers provoked metropolitan England into a theory of direct control rather than influence through suggestion and bribery. Annexation was an answer to the decline in British industrial supremacy and sea-power.

19. Governor Thomas Ludlam to Zachary Macaulay, 14 April, 1807, quoted in Philip D. Curtin, *The Image of Africa: British Ideas and Action, 1780–1850* (London, 1965), p. 255.

20. See Joseph Conrad, *Heart of Darkness*. The text is taken from his *Collected Works* (London, 1925).

21. A brilliant review, 'Painting the Unpaintable', by Richard Dorment in the *New York Review of Books*, 27 September, 1990, refutes the views of Guy C. McElroy and Albert Boime on the Black Image in the art of the nineteenth century.

22. *The Frontier of History: North America and Southern Africa Compared* (H. Lamar and L. Thompson eds., New Haven, 1982) is a seminal work of comparative history. It was followed by an important review, 'Settlers and "Savages" on Two Frontiers' by George M. Fredrickson, *New York Review of Books*, 18 March, 1982.

5
THE MYTH OF THE SAVAGE

1. See the Penguin Classics version of *The Epic of Gilgamesh*, p. 39.

2. Again I have used the Penguin Classics version of Sir Thomas More's *Utopia* (London, 1965). Zavala wrote an excellent monograph on *Sir Thomas More in the New Spain*.

3. See L. Hanke, *Aristotle and the American Indian*, pp. 78–9.

4. As well as the work of Howard Mumford Jones and Hanke, there is an excellent chapter on this theme called 'In Utopia and among the Savages' in Elizabeth Rawson's *The Spartan Tradition in European Thought* (Oxford, 1969), pp. 170–85. The best general treatment is to be found in H.N. Fairchild, *The Noble Savage* (New York, 1928).

5. The two contemporary accounts of Sir Thomas Gates's shipwreck are to be found in William Strachey's 'A True Repertory of the Wrack and Redemption of Sir Thomas Gates ... upon and from the Islands of the Bermudas ...' Although it circulated in manuscript form, it was first printed in Purchas, *Pilgrims* (London, 1625), IV, pp. 1734–58. It has been reprinted by Louis B. Wright in *A Voyage to Virginia in 1609: Two Narratives* (Charlottesville, 1964), as has the account from which this quotation is taken, Silvester Jourdain, *A Discovery of the Bermudas, Otherwise Called the Isle of Devils* (London, 1610).

While there is no proof that Shakespeare read either of these accounts or other accounts of voyages to Virginia, the whole plot of *The Tempest*, the frequent references to Caliban as an Indian (cf. *Stephano*: 'What's the matter? Have we devils here? Do you put tricks upon's with savages, and men of Ind, ha?'), and finally Ariel's speech about the location of the wrecked ship make it likely that he knew of the occasion.

> *Ariel*: Safely in harbour
>   Is the king's ship, in the deep nook, where once
>   Thou call'dst me up at midnight to fetch dew
>   From the still-vex'd Bermoothes, there she's hid . . .

Bermoothes was another name for Bermudas, and Gates's ship did, indeed, 'fall in between two rocks, where she was fast lodged and locked for further budging.'

6. I have used the New Temple Shakespeare edition for the quotations from *The Tempest* and *Titus Andronicus*.

7. See *A Letter Sent into England from the Summer Islands. Written by Mr Lewis Hughes, Preacher of God's Word There* (London, 1615), reprinted in *The Elizabethan America*, L. Wright ed., pp. 202–5.

8. Thomas Hobbes, *Leviathan, or the Matter, Form and Power of a Commonwealth, Ecclesiastical and Civil* (M. Oakeshott ed., Oxford, 1955), pp. 82–3.

9. Thomas Jefferson to the Marquis de Chastellux, 7 June, 1985, from *The Papers of Thomas Jefferson* (J. Boyd ed., Princeton, 1953), VIII, p. 185. It is interesting to see that Jefferson from slave-owning Virginia could not extend the vigorous qualities of the New World immediately to the black man. The letter continues: 'I believe the Indian then to be in body and mind equal to the white man. I have supposed the black man, in his present state, might not be so. But it would be hazardous to affirm that, equally cultivated for a few generations, he would not become so.'

10. See Lewis O. Saum, *The Fur Trader and the Indian* (Univ. of Washington, 1965), *passim*. This important book was the first to collect and analyse the account of the first businessmen to deal with the Indians of the forests. It was followed by many revisionist works supporting the Indian position from the point of view of ecology. Philip G. Terrie, 'Recent Work in Environmental History', and Brian W. Dippie, 'American Wests: Historiographical Perspectives', *American Studies International*, October, 1989, Vol. XXVII, No. 2, give admirable summaries of the new perspectives on the history of the west in the United States. Particularly important are Richard C. Wade, *The Urban Frontier: The Rise of Western Cities, 1790–1830* (Cambridge, Massachusetts, 1959), which stated the towns were the spearheads of the frontier; Earl Pomeroy, 'Toward a Reorientation of Western History: Continuity and Environment', *Mississippi Valley Historical Review*, No. 41, March, 1955; Robert G. Athearn, *The Mythic West in Twentieth-Century America* (Lawrence, Kansas, 1986); Calvin Martin, *Keepers of the Game: Indian-Animal Relationships and the Fur Trade* (Berkeley, California, 1978); *Indians, Animals and The Fur Trade* (S.

NOTES

Krech III ed., Athens, Georgia, 1981); Paul Radin, *The Autobiography of a Winnebago Indian* (Univ. of California, 1920); Charles L. Sanford, *The Quest for Paradise and the American Moral Imagination* (Urbana, Illinois, 1964); Roderick Nash, *Wilderness and the American Mind* (rev. ed., New Haven, Connecticut, 1973); and Richard Slotkin in *Regeneration Through Violence: The Mytholoy of the American Frontier, 1600–1860* (Middletown, Connecticut, 1973) and *The Fatal Environment: The Myth of the Frontier in the Age of Industrialization, 1800–1890* (New York, 1985); and *This Well-Wooded Land: Americans and their Forests from Colonial Times to the Present* (T. Cox, R. Maxwell, P. Thomas, J. Malone, eds., Lincoln, Nebraska, 1985).

11. The quotation is from a sympathetic enquiry into the Indian attitude to their environment by J. Donald Hughes, *American Indian Ecology* (El Paso, Texas, 1983).

12. Luther Standing Bear, *Land of the Spotted Eagle* (Lincoln, Nebraska, 1978).

13. *Black Elk Speaks* (as told through John G. Neihardt, New York, 1932).

14. This extract is taken from a modern edition of Smith's *General History* called *Captain John Smith's America* (J. Lankford ed., New York, 1967), p. 26.

15. See Edward Johnson, *Wonder-Working Providence* (Boston, 1653), and *Narratives of the Indian Wars, 1675–1699* (Charles Lincoln ed., New York, 1913), p. 105. Although most contemporary accounts condemned all Indians as evil, Daniel Gookin in his *Historical Collections of the Indians in New England*, (Boston, 1677) defended the few Indians who became Christians.

16. See John Williams, *The Redeemed Captive Returning to Zion* (Boston, 1707), p. 25. In his *Regeneration Through Violence: The Mythology of the American Frontier*, Richard Slotkin analysed the captivity myth in such accounts, and the Puritans' need to justify themselves both as passive victims and bloodthirsty avengers.

17. See James Seaver, *A Narrative of the life of Mrs Mary Jemison* (new ed., New York, 1961).

18. The most illuminating commentary on the Leatherstocking saga, which sets it brilliantly in its social and historical context, is to be found in Henry Nash Smith's *Virgin Land: The American West as Symbol and Myth*.

19. For the text of *The Scarlet Letter*, I have used *The Collected Works of Nathaniel Hawthorne* (New York, 1939).

20. Timothy Flint published the *Western Monthly Review* between 1827 and 1830 and tried to urge pioneers from the east to go west. This extract comes from his *Indian Wars of the West: Containing Biographical Sketches of Those Pioneers who Headed the Western Settlers in Repelling the Attacks of the Savages* (Cincinnati, 1833). His novel *The Shoshonee Valley*, published in 1830, was the first to use Mountain Men as characters. Flint's opinion of them and the fur trappers was low. They had 'an instinctive fondness for the reckless savage life, alternately indolent and laborious, full and fasting, occupied in hunting, fighting, feasting, intriguing, and amours, interdicted by no laws, or difficult

morals, or any restraints but the invisible ones of Indian habit and opinion.'

21. I have taken the quotations from J. Donald Crowley's admirable edition of *Robinson Crusoe* (Oxford, 1972), pp. 113, 171, 146.

22. See D. Hay, *Europe*, pp. 104, 120–21. This book remains the best on the emergence of Europe as an idea.

23. From G.A. Montecuccolo Cavazzi, *Istoria descriptione de tre regni Congo, Matamba et Angola* (Rome, 1687). Quoted in B. Davidson, *Black Mother*, pp. 32, 103.

24. S.M.X. Golberry (de Golbery), *Travels in Africa performed in the Years 1785, 1786 and 1787* (2 vols., London, 1803), II, p. 240.

25. Dr B. Mosley, *A Treatise on Tropical Disease; and on the Climate of the West Indies* (London, 1787), p. 48.

26. David Hume, 'Of National Character', from *The Philosophical Works of David Hume* (4 vols., London, 1898), III, p. 252. This is quoted by Curtin, *The Image of Africa*, p. 42. I am indebted throughout this section of the book to Professor Curtin's insights and researches on the influence of Africa in Britain between 1750 and 1850.

27. See James Boswell's *Life of Johnson* (Everyman ed., Vol. I), p. 271.

28. *The Novels of Aphra Behn* (E. Baker ed., London, 1905), p. 41. See D. Grant *The Fortunate Slave*, for the sympathetic treatment given to the Gambian, Job ben Solomon, after his redemption from slavery in Maryland.

29. A declaration of 1774 by Charles Wheeler, quoted in B. Davidson, *Black Mother*, p. 102.

30. I am indebted to the opening chapter on the Jesuits in China, to be found in J. Spence's *To Change China: Western Advisers in China, 1620–1960* (Boston, 1969), from which this quotation from Ricci is derived.

31. Chang Hsieh in the *Tung Hsi Yang Khao* of 1618.

32. Michel de Montaigne, 'Of Coaches' from *The Complete Works of Montaigne: Essays, Travel Journal, Letters* (D. Frame ed., Stanford, 1957), pp. 694–5.

33. Mary Midgley, *Beast and Man*, is excellent on the subject of Descartes and his use of reason to divorce the human species from all others.

34. This was Horace Walpole's phrase about the landscaping of William Kent which improved nature.

35. Alexander Pope's poems come from his *Moral Essays* and *Of Taste*.

36. Donald Worster, *Nature's Economy: A History of Ecological Ideas* (Cambridge, 1985) is excellent on the importance of Gilbert White.

37. Leo Marx, *The Machine in the Garden: Technology and the Pastoral Ideal in America* (Oxford, 1964) is seminal on the subject in England and in the United States.

NOTES

38. The quotation is from the Earl of Shaftesbury's *Characteristics* of 1711. For understanding the penal code of the Age of Reason, essential reading is *Albion's Fatal Tree: Crime and Society in Eighteenth-Century England* (D. Hay, P. Linebaugh, J.G. Rule, E.P. Thompson, C. Winslow, London, 1975).

39. Curiously enough, the British government's Riot Acts played the same role as the *Requerimiento* in turning resistance into rebellion. A mob of machine-breakers or of farm labourers demanding 'bread or blood' could be declared traitors by the reading of unintelligible words at a distance. Only at 'Peterloo' in 1819, when the local yeomanry attacked the reformist crowds at Saint Peter's Field in Manchester before the formal reading of the Riot Act, did a storm of protest arise. Some of the eleven dead citizens seemed to have been murdered, rather than been found dead after the legal scattering of a mob. Yet even so, the magistrates and the military were thanked by the government for their 'prompt, decisive and efficient measures for the preservation of public tranquility.' At the last resort, the niceties of the Act bowed to the keeping of the peace. Rule was more important than law.

40. While the roots of the Romantic revolt are often ascribed to William Wordsworth and his associated group of poets, their influence seems only conservative and national.

41. From Canto VIII of *Don Juan*, LXI-LXVII.

6
THE WILD VERSUS THE MACHINE

1. I have used the texts from *The Novels of Thomas Love Peacock* (David Garnett ed., London, 1948).

2. E.S. Turner, *All Heaven in a Rage* (London, 1964) quotes Lord Erskine's bill of 1809 and remains definitive on animal rights and the role of the Royal Society for the Prevention of Cruelty to Animals.

3. Emerson is quoted in the essential book on this theme, R.W.B. Lewis, *The American Adam: Innocence, Tragedy and Tradition in the Nineteenth Century* (Chicago, 1955).

4. David W. Noble is particularly good on the writings of James Fenimore Cooper and Hawthorne and Melville in *The Eternal Adam and the New World Garden* (New York, 1968).

5. Barbara Novak, *Nature and Culture: American Landscape and Painting, 1825–1875* (London, 1980) makes a significant contribution to the study of American perceptions in the wilderness. Also illuminating is Joshua C. Taylor, *America As Art* (New York, 1976).

6. Joseph Kastner, *A Species of Eternity* (New York, 1977) is extremely useful on the beginnings of natural sciences in the United States and particularly on Peale's Museum.

7. These quotations from H.B. Stowe, *Dred, A Tale of the Dismal Swamp* (2 vols., Boston, 1856) and T.W. Higginson, 'Barbarism and Civilisation', *Out-Door Papers* (Boston, 1863) were brought to my notice by the significant study of David C. Miller, *Dark Eden: The Swamp in Nineteenth-Century American Culture* (Cambridge, 1989). Also useful are Christopher Mulvey, *Anglo-American Landscapes: A Study of Nineteenth-Century Anglo-American Travel Literature* (Cambridge, 1983) and *Views of American Landscapes* (M. Gidley and R. Lawson-Peebles eds., Cambridge, 1989).

8. Professor Curtin's chapter on 'The Racists' in *The Image of Africa* is excellent on the rise of racist thought in the 1850s. Also useful is the recent biography of the Count de Gobineau by Michael D. Biddiss, *Father of Racist Ideology* (London, 1970). The quotation is to be found in *Gobineau: Selected Political Writings* (M. Biddiss ed.), p. 163. Gobineau's friend de Tocqueville proposed one of the better criticisms of the theory of race, when he wrote of the *Essays on the Inequality* ...: 'What advantage can there be in persuading base peoples living in barbarism, indolence or slavery that, such being their racial nature, they can do nothing to improve their situation or to change their habits and government? Do you not see inherent in your doctrine all the evils engendered by permanent inequality – pride, violence, scorn of fellow men, tyranny and abjection in all their forms?'

9. Lewis Mumford, *The Myth of the Machine* (New York, 1966), p. 240, provides this analysis of the technological city, one of many extraordinary insights in an important work.

10. See Jack London, *People of the Abyss* (London, 1903), *passim*. Along with Gorki's autobiography and George Orwell's *Down and Out in Paris and London*, Jack London's work remains the best account by a novelist of the degradation of the poor under modern industrial conditions.

11. Marx to Engels, 18 June, 1862, from *Marx–Engels Selected Correspondence* (London, 1956), p. 157.

12. Charles Darwin, *The Descent of Man* (rev. ed., New York, 1890), p. 619.

13. Edward Burnett Tylor, *Anthropology: An Introduction to the Study of Man and Civilisation* (London, 1881), p. 24.

14. Herrera in his *General History*, II, p. 16, even asserted anthropological evidence for the degenerative theory of the savage by telling of an old Indian in Cuba who had claimed knowledge of a Flood and the Ark 'with other particulars, as far as Noah's sons covering him when drunk, and the other scoffing at it; adding, that the Indians descended from the latter, and therefore had no coats nor cloaks; but that the Spaniards, descending from the others that covered him, were therefore clothed and had horses.'

15. A useful introduction to palaeontology can be found in Herbert Wendt, *Before the Deluge* (New York, 1968), and to the history of social anthropology in H.R. Hays, *From Ape to Angel* (New York, 1958). The works of the authorities mentioned in the text are readily available. The specialist on the extinction of

NOTES

prehistoric species was Wolfgang Soergel, who was primarily occupied with the Ice Age.

16. From Lautréamont's *Poésies* (Paris, 1870).

17. From an appreciation of Gauguin's art, Charles Estienne's *Gauguin* in the excellent Skira series still remains important.

18. *Journal of the Right Hon. Sir Joseph Banks during Captain Cook's First Voyage in H.M.S. Endeavour* (J. Hooker ed., London, 1896), p. 74. The full version of Banks's journal was edited later by J.C. Beaglehole and published in 1962. See also the Canberra manuscript of Banks, *Thoughts on the Manner of Otaheite* written for the amusement of the Prince of Orange in 1773. I am indebted to the important book by Bernard Smith, *European Vision and the South Pacific, 1768–1850* (Oxford, 1960), for this important reference and for its scholarly groundwork in the field more racily covered by Alan Moorehead in *The Fatal Impact* (London, 1966).

19. See Louis Antoine de Bougainville, *Voyage round the World* (trans. J.R. Forster, London, 1772), pp. 228–52.

20. Daisy Bates, *The Passing of the Aborigines: A Lifetime Spent among the Natives of Australia* (London, 1938), p. 87.

21. Quoted in Dean Boyce's excellent monograph *Clarke of the Kindur: Convict, Bushranger, Explorer* (Melbourne, 1970), p. 66.

7
THE GREAT ENGINEER

1. The quotation is from Richard Rhodes's penetrating book, *The Inland Ground: An Evocation of the American Middle West* (New York, 1970), pp. 11–13.

2. John Locke's remarks on property in America appear in 'An Essay Concerning the True Original, Extent and End of Civil Government' (London, 1690).

3. Mark Twain, *Life on the Mississippi* (New America Library edn., 1961), pp. 93–95.

4. For the ambiguity of Emerson and Thoreau about the progress of the railroad through the wilderness, I am indebted to the seminal work of Leo Marx, *The Machine in the Garden*, op. cit., *passim*. See also John R. Stilgoe, *Metropolitan Corridor: Railroads and the American Scene* (New Haven, Connecticut, 1983).

5. In his analysis of the 'megamachine' of civilisation, Lewis Mumford lists these chief mechanical agents in *The Myth of the Machine*, op. cit., pp. 286–287.

6. See the concluding paragraphs of Frank Norris, *The Octopus*, New York, 1901.

7. This opinion of the Cattle Commissioners of New York in 1869 is quoted in E.S. Turner, *All Heaven in a Rage*, op. cit.

8. David M. Potter, *People of Plenty: Economic Abundance and the American*

Character (Chicago, 1954) remains a seminal work on its subject.

9. Andrew Ure, *The Philosophy of Manufacturers*, was published in London in 1835 and Thomas Ewbank, *The World a Workshop, or, The Physical Relationship of Man to the Earth* in New York in 1855. I am again grateful to D. Worster, *Nature's Economy*, op. cit., for bringing Ewbank's book to my attention.

10. Of the many accounts of the Zulu War, I have found the most concise and balanced to be R. Furneaux, *The Zulu War* (London, 1963).

11. The biography of H.M. Stanley by Richard Hall gives a convincing psychological portrait of the adventurer, although it minimises his taste for slaughter. Still important for an understanding of the conditions of the Congo under King Leopold are the Casement report of 1904 and E.D. Morel's crusading attack on *King Leopold's Rule in Africa* (London, 1907).

12. See B. Davidson, *Black Mother*, pp. 203–4.

13. Commander R.H. Bacon, *Benin, the City of Blood* (London, 1897), *passim*. Bacon seems like many Europeans to have been less overawed by the sights than the smells, which he reports 'no white man's internal economy could stand'.

14. See D. Livingstone, *Private Journals, 1851–1853*, *passim*.

15. Brian V. Street, *The Savage in Literature: Representations of 'primitive' society in English fiction 1858–1920* (London, 1975) is too narrow in its consideration of the meaning of the word 'savage', although its theme is important. Also useful are R.H. Pearce, *The Savages of America* (Baltimore, 1965); and L.J. Henkin, *Darwinism in the English Novel 1860–1910* (New York, 1940).

8
THE CULT OF THE SAVAGE

1. See Horace Walpole, *Journals of Visits to Country Seats* (Walpole Society, 1928), XVI, pp. 9–80 *passim*.

2. D. Defoe, *A Tour thro' the Whole Island of Great Britain* ... (London, 1742), III, p. 231.

3. D.C. Webb, *Observations and Remarks During Four Excursions Made to Various Parts of Great Britain in the Years 1810 and 1811* (London, 1812), pp. 189–90.

4. Mrs Radcliffe, *A Journey Made in the Summer of 1794 ... with Observations During a Tour of the Lakes in Westmoreland and Cumberland* (London, 1795), p. 465.

5. Esther Moir in *The Discovery of Britain: The English Tourists 1540 to 1840* (London, 1964) is excellent on the change of the perceptions of travellers from the praise of national progress to the thrill of romantic excess.

6. *The Complete Works of Ralph Waldo Emerson* (E.W. Emerson ed., Boston,

NOTES

1903–4), I, p. 9. Morton and Lucia White, *The Intellectual versus the City: From Thomas Jefferson to Frank Lloyd Wright* (Boston, 1962) remains the most important work on the anti-urban sentiments of American intellectuals.

7. This phrase of Parkman is quoted in Henry Nash Smith in his important *Virgin Land: The American West as Symbol and Myth* (Boston, 1950), while dealing with the subject of 'Daniel Boone: Empire Builder or Philosopher of Primitivism?'

8. See Carl G. Jung, *Man and His Symbols* (New York, 1964).

9. Both R.W.B. Lewis, *The American Adam: Innocence, Tragedy and Tradition in the Nineteenth Century* (Chicago, 1955) and David W. Noble, *The Eternal Adam and the New World Garden: The Central Myth in the American Novel since 1830* (New York, 1971) have provided many insights into the theme of American innocence.

10. Frederick Jackson Turner, 'The West and American Ideals', *Washington Historical Quarterly*, October, 1914, V, p. 245.

11. Theodore Dreiser, *Sister Carrie* (New York, 1900) p. 70.

12. In one striking image in *The People of the Abyss* (London, 1903), Jack London wrote of the East End of London as if it were an infuriated tigress turning on its young.

13. Jack London to Anna Strunsky, 31 July, 1902, *Huntington Library*.

14. J. London, *The People of the Abyss*, pp. 283–5.

15. Ibid, p. 320.

16. Upton Sinclair, *The Jungle* (New York, 1906), pp. 197, 270.

17. Oscar Handlin, *Race and Nationality in American Life* (Boston, 1957), pp. 157, 166.

18. Hannah Arendt, *The Burden of Our Time* (London, 1951), p. 266.

19. Ibid, pp. 295–6.

20. The quotation is taken from Hanna Hofkesbrink, *Unknown Germany* (New Haven, 1948), p. 21.

21. H. Arendt, *The Burden of Our Time* (London, 1951), p. 266.

22. Both of Malinowski's important works on the Trobriand Islanders have been reissued and are in print, *Crime and Custom in Savage Society* (London, 1926) and *Sex and Repression in Savage Society* (New York, 1927).

23. André Malraux, *The Voices of Silence* (New York, 1955), p. 352.

24. *Istorica descrizione de tre regni Congo* ..., op. cit., p. 103.

25. D.H. Lawrence, *Studies in Classic American Literature* (Anchor ed., New York, 1953), pp. 43–4.

9
THE RESPECTABLE SAVAGE

1. See the preface to *Crime and Custom in Savage Society*, op. cit., p. xi.
2. Joyce Cary, *Mister Johnson* (London, 1950), pp. 207, 212.
3. Gregory Bateson, quoted in S. Brand, *II Cybernetic Frontiers* (New York, 1974), p. 33.
4. See Hugh M. Hole, *The Passing of the Black Kings* (London, 1932), pp. 175, 231.
5. Edward O. Wilson, 'Human Decency is Animal', *New York Times*, 12 October 1975.
6. Edward O. Wilson, *Sociobiology: A New Synthesis* (Boston, 1975), p. 547 and *passim*. For a searching critique of Wilson's theories of 'The Morality of the Gene' as tending towards determinism and totalitarianism, see M. Midgley, *Beast and Man*, *passim*.
7. See P. Amaury Talbot, *In the Shadow of the Bush* (London, 1912), as quoted by Adolf E. Jensen, *Myth and Culture among Primitive Peoples* (Chicago, 1963), pp. 25–9.
8. In his book, *The Mind of Man in Africa* (London, 1972), J.C. Carothers presents a daring thesis, based on McLuhan's insights, that the 'ear culture' of black Africa has led to its misunderstandings with the 'eye culture' of Europeans. I do not follow the author, however, in his reliance on the work of C.S. Coon and W. Howells, when they seek the origins of the 'races' of humanity.
9. J. Conrad, *Heart of Darkness*, pp. 95–6. *Horror sylvanus* is still a rare psychiatric disease found among Germans who live deep in the pine forests.
10. Sir Frank Fraser Darling, quoted in *The Observer*, 8 September, 1974.
11. Quoted in *The Times*, 20 July, 1975.
12. Quoted in the *New Statesman*, 4 January, 1974.
13. Gilberto Freyre, *The Masters and the Slaves: A Study in the Development of Brazilian Civilisation* (2nd rev. ed., New York, 1968), I, p. xxix.
14. Ibid, I, p. 5.
15. See E.O. Wilson, *Sociobiology*, p. 250.
16. Peter Wilsher's article 'Breaking Point' in the *Sunday Times* of 31 March, 1974, has been most useful in my consideration of the effect of the overpopulation of the cities of the Third World.
17. Peter Matthiessen in *Under the Mountain Wall: A Chronicle of Two Seasons in the Stone Age* (New York, 1962) has given a perceptive description of the behaviour of the last of the New Guinea tribes as they met white people for the first time.

18. For the documentation on the new export of torture techniques disguised as foreign aid, see the anonymous author of 'Building a Better Thumbscrew', *New Scientist*, 19 July, 1973.

19. H. Brackenridge as quoted in *Indian Atrocities: Narratives of the Perils and Sufferings of Dr Knight and John Slover, among the Indians ...* (Cincinnati, 1867), p. 62.

20. *New York Times*, 10 February, 1964.

10
GREEN AND DYING

1. Alexis de Tocqueville, *Democracy in America* (P. Bradley ed., New York, 1945), I, pp. 22–6.

2. Aldous Huxley, *Brave New World* (Penguin new ed., London, 1946), p. 187.

3. Ibid, preface, p. 7.

4. Aldous Huxley, *Brave New World Revisited* (Bantam ed., New York, 1960), pp. 6, 22, 64.

5. See E. Callenbach, *Ecotopia* (London, 1978).

6. Frantz Fanon, *Toward the African Revolution* (London, 1970), pp. 190–191.

7. J.S. Weiner, a professor of Environmental Physiology, as quoted by Timothy Severin in his useful *Vanishing Primitive Man* (London, 1973).

8. Colin M. Turnbull is the author of the important *The Forest People* (London, 1961) and *The Mountain People* (London, 1966), also of *Man in Africa* (New York, 1976).

9. See J.J. Rousseau, *A Discourse upon the Origin and Foundation of the Inequality among Mankind*.

10. See John Updike's introduction to *Soundings in Satanism* (New York, 1975).

11. The Australian physicist Mark Oliphant made this remark on the dropping of the first atomic bombs, as quoted in this author's *The Red and the Blue* (London, 1986), p. 115.

12. No summary can do justice to Rachel Carson, *Silent Spring* (New York, 1962).

13. See James Lovelock, *The Ages of Gaia: A Biography of Our Living Earth* (New York, 1988).

14. Desmond Morris, *The Naked Ape: A Zoologist's Study of the Human Animal* (London, 1967), is fascinating, but excessive, on the correspondences between human and animal behaviour patterns.

15. See *The Times*, 25 September, 1990. Also see Peter Singer, *Animal Liberation* (New York, 1990).

16. The Duke of Edinburgh, speaking as the President of the World Wildlife Fund for Nature (formerly the World Wildlife Fund) in Washington, D.C., as reported in *The Times*, 19 May, 1990.

17. See T. Dale and V.G. Carter, *Topsoil and Civilisation* (Univ. of Oklahoma, 1955). It is also quoted in the important economic work of E.F. Schumacher, *Small Is Beautiful: A Study of Economics as if People Mattered* (London, 1973).

18. Edward Carpenter is quoted in 'The World is Dying: What Are You Going To Do About It?', *Sunday Times Magazine*, 26 February, 1989, one of many media presentations which demonstrated a popular concern for the pollution of the planet.

19. Most important in recent popular ecological works have been in the United States, Bill McKibben, *The End of Nature* (New York, 1990), and Wendell Berry, *What Are People For?* (North Point, 1990) and in Britain, Richard Mabey, *Home Country* (London, 1990).

INDEX

aborigines, 118–20, 193
  animal paintings, 120
  dispossession, 172
  Europeans living with, 119
acid rain, 173
Adam, 108
  in American thought, 195
  cult of innocent, 108, 109
  and Eve, 18, 152
Africa
  ambiguous effect of communications and civilization, 145
  British interest, 68–9
  Christian missions, 65, 95
  climate, 94
  colonisation, 186–7
  conditions before European arrival, 60
  European exploration to centre, 65–6
  European occupation, 73
  European superiority feelings, 93
  European voyages to, 28
  exploitation, 68
  influence on Britain, 190
  influence on Europe before eighteenth century, 92–3
  new national governments, 146
  perceptions of backwardness, 93–5
  sharing out between white powers, 69
  sophistication of people taken as slaves, 61
Africa and the Victorians, 186
Africans
  cult of Noble Savage, 94
  displacement by colonists, 74
  explorers' attitudes towards, 66–7
  as savages, 94
Age of Enlightenment, 83, 94
Age of Reason, 40, 99, 191
  cruelty in, 100
  view of primitive being, 106
Agent Orange, 157
Ages of Gaia, The: A Biography of Our Living Earth, 197
aggression
  and overcrowding, 158
  recognition in men, 90
  in urban life, 163
agriculture
  civilised man against savages, 121
  factory farming methods, 169
aid programmes, 159, 197
Ainus of Yesso, 2, 176
Alexander of Macedon, 1, 12
All Heaven in a Rage, 178, 191
Altamira (Spain), 1
altruism, 150
Amazon rain forest, 155
  clearance, 155
  liquidation of Indians, 155–6
America
  Buffon's view of weakness of living things, 84
  comparison with Europe, 84
  conflicts in, 161
  cult of innocent Adam, 108, 109
  cultures of rich and poor in cities, 161
  development of railways, 122, 123, 193
  economics for settlers, 122
  industrial civilisation, 53
  Jefferson's view of vigour, 84
  Montaigne's views, 98
  nineteenth century artists, 108–9
  perversion of realities for European purposes, 84
  realist school of novelists, 135
  state of war in, 83

INDEX

America – *contd.*
   technology, 122
   trade development, 122
   transport, 122
   urban man, 161
   wilderness, 133
American Adam, The: Innocence, Tragedy and Tradition in the Nineteenth Century, 191
American Dream, An, 160
American Indians, 2
   attacks on colonists, 48
   colonists' opinions of, 47, 48, 49, 51
   conflict of culture meeting Europeans, 182
   de Tocqueville's perceptions, 163, 164
   descriptions by early traders, 85
   displacement for Caribbean, 42
   dispossession of, 74, 82, 86
   Dominicans' protests at maltreatment of, 33
   early opinions of Spanish, 40, 41
   ending of harmony with nature, 86
   enslavement by Spanish, 42
   and environment, 86
   Harriot's account, 47
   in Icelandic sagas, 37–9
   Lawrence's views on, 144
   metal working, 44
   nakedness, 41
   novelists' inspiration, 88
   policies of extermination, 90
   prisoners' accounts, 87–8
   records of encounters with Europeans, 36–7
   recovery of lands, 172
   roles for colonists, 48
   Rousseau's views on, 85
   Smith's observations, 87
   Spanish debate, 29, 32
   trade with, 188
   trapping, 86
   view of nineteenth century artists, 109
   war with, 83, 87
Anatomy of a Pygmie compared with that of a Monkey, an Ape, and a Man, 64
Ancient Society, 115
Andaman Islanders, 142
Animal Liberation, 169
animals
   aborigine paintings, 120
   abuse, 108, 124
   attitudes to in Age of Reason, 100

   baiting, 31, 169
   Buddhist attitudes to, 23
   Christian attitudes to, 19–20
   legal rights, 107
   Marx's views, 111–12
   moralist's fear of in man's nature, 55
   paintings, 170
   as religious images, 31
   Roman attitude to, 13–14
   Romantics' view of relationship with humans, 107
   as sport, 169
   worship, 120, 177
anthropologists living with tribes, 143
anthropology, 113–14
   and colonial powers, 145
   Darwinian, 115
   modern, 129
   studies of last primitive peoples, 141
   Tylor's concepts, 176
Anthropology: An Introduction to the Study of Man and Civilisation, 113, 192
apartheid, 148
apocalypse, 139
Apologetic History, 34
Arab civilisation, 21
   destruction by Turks and Mongols, 22
   survival, 23
Aral Sea, 159
archaeology, 176–7
Ardrey, Robert, 150
Arendt, Hannah, 138, 140, 195
Aristotle, 1, 11, 13, 23, 34, 149, 178
   ostracism, 36
Aristotle and the American Indians, 181
art
   decadence in Victorian age, 117
   development, 1
   perceptions by different societies, 142
artists, 193
   aborigine, 120
   American nineteenth century, 108–9
   return to primitive simplicity, 117, 118
Aryan race, 149
Ashanti, 125
   British expedition against, 69
   slaves of, 59
Asia, new national governments, 146
Assassins, 76
assimilation philosophy, 148
astronomy
   Chinese, 97
   development in desert regions, 154

INDEX

Atala, 88
Athens, 10–11
Atlantis, 75
atmosphere stabilisation, 171
atomic
 age, 168
 weapons, 197
Australia
 European refugees, 119
 fate of aborigines, 118–20
 penal colonies, 119–20
authoritarian rule, 83
automobiles, 123
Aztecs, 44
 collapse of empire, 46
 fear of trained beasts, 44–5

Bacchae, 12
Back to the Land movement, 171
Baker, Sir Samuel, 66–7, 186
Ballantyne, R M, 129
Bancroft, George, 132
Banks, Sir Joseph, 117, 193
Bantu homelands, 147
barbarians, 11–12, 13
 Roman attitude to, 14–15
 sloughing off civilisation, 160
barbarism
 Cambodia, 147
 under Third Reich, 140
barbed wire, 123, 126
Barbot, John, 186
Bates, Daisy, 118, 119, 193
Bateson, Gregory, 146
bear festivals of Ainu, 176
Beast and Man, 190
beauty, enjoyment of, 142
beavers, 54
behaviour
 anti-social, 150–1
 human, 169
behaviourism, 149
Benin (formerly Dahomey), 128, 194
Benn, Aphra, 94–5
Bible, 24, 25
Big House, 157
Billington, Ray Allen, 184–5
Bilston ironworks, 130–1
biological control, 168–9
biological weapons, 160
biology in human development, 149
Black Elk Speaks, 86, 189
Black Mother, 190, 194

Black Power, 61
blacks
 barbarism in ghettoes, 161
 dance, 143
 effects of industrial system of division from whites, 111
 perceptions of inferiority to whites, 94
Blake, William, 107
Boas, Franz, 142
Boers, 126, 147
 justification for segregation, 148
Boller, Henry, 85
Borrowdale Gorge, 131
Boswell, James, 59
Bougainville, Louis Antoine de, 118
Boyle, Robert, 99
Brackenridge, Hugh Henry, 161
Bradford, William, 49, 50, 184
Brasília, 155
Braudel, Fernand, 178
Brave New World, 143, 151, 164–5, 166, 197
Brave New World Revisited, 165, 197
Brazil
 land clearance, 155
 slavery by Portuguese, 157
Brief and True Report of the New-Found Land of Virginia, 183
Broken Spears, The, 182, 183
Brown, Claude, 161
Buckley, William, 119
Buddhism, 23
 attitudes to nature, 23–4
buffalo, 52–3
Buffon, Comte de, 84
Burden of Our Time, The, 195
Burton, Sir Richard, 67, 186
Byron, Lord George Gordon, 101, 102, 104–5

Caitlin, George, 109
Calcutta, 158
Caleb Williams, 101–2
Caliban, 24, 25, 78–9, 80
Callenbach E, 197
Cambodia, 147
Camino Real, 173
Camoëns, 42
Campaign for Nuclear Disarmament, 168
cannibalism, 118, 128, 183
capitalism
 laissez-faire, 134
 Marx's views, 111

INDEX

capitalists, 61
Captain John Smith's America, 189
captivity myth, 189
Carpenter, Edward, 171, 198
Carson, Rachel, 154, 168, 197
carving, rock, 1
Cary, Joyce, 145, 196
Casement, Sir Roger, 127, 155
casual labourers, 134, 137
catastrophic creation theory, 114–15
Catholic faith, 32
Catholicism
  in Brazil, 157
  Spaniards' conversion of Indians to, 90
cattle, 123, 124
Cattle Commissioners of New York, 124, 193
Cavazzi, Father, 93
cave paintings, 116
cave-bears, 116
Celts, 160
  tree worship, 17
Ceuta, conquest of, 28
Charles, Prince of Wales, 170
Charles the Second, King of England, 82–3
Chateaubriand, François René, 88
Chelmsford, Lord, 125
Chernobyl, 171
Chestnut, Mary B, 186
Chicago, 136–7
Ch'in dynasty (China), 8
China, 8–9
  cities, 8
  control of beast within in medieval time, 55
  counter-civilisation to Europe, 97
  culture, 8
  descriptions of civilisation, 96
  frontier theory in history, 177
  Great Wall, 55, 98
  Jesuits in, 96–7, 190
  Mongol conquest, 55
  rejection of west, 97–8
  self-sufficiency, 56
  slave-trade expeditions, 179
China: A Short Cultural History, 177
chinoiserie, 97
Chomsky, Noam, 26, 151–2, 180
Chou dynasty (China), 8
Christian and his Comrades, 105
Christianity
  attitudes to animals, 19–20
  attitudes towards Africa of Europeans, 93
  de Gobineau's views, 111
  definition of moral practice, 16
  equality of souls doctrine, 23
  humility, 95
  identification with Jerusalem, 16
  ideology of innate difference and superiority, 24
  man's relationship with nature, 170
  missionaries, 65, 95, 156
  myth of creation, 17–18
  religious justification for war, 45–6
  rise of, 15
  saints, 19–20
  Second Coming, 172
  spread, 16
  symbolism of trees, 152–3
Christianopolis, 78
Cicero, 14
cities
  of China, 8
  crime rate, 160
  culture, 4–5
  cultures of rich and poor, 161
  economic mismanagement, 136
  overpopulation, 196
  regrowth of woods round, 162
  relationship of savage with, 7–8
  savagery in, 162, 164
  technological, 192
  Third World, 157–8
  as vicious jungle, 135
City of God, 16
Civil War (America), 110
civilisation, 115
  degeneracy, 116–17
  dispossession of primitive people, 164
  fall, 117
  materialistic, 172–3
  nature of men in relation to, 149
  perception of savage, 167
  pleasure-controlled, 165
  Victorian belief in superiority, 168
Civitas Solis, 78
Clarke, George, 119
Clarke, Marcus, 120
Clockwork Orange, A, 162, 164
clothes, 1
coal mining, 111
colonial administration, 146
colonial uprisings, 61

INDEX

colonialism, 29
  Shakespeare's representation, 79
colonists
  barriers as fears, 54
  crop growing, 47–8
  displacement of Indians, 53
  Hakluyt's instructions for, 47–8
  savagery amongst Europeans, 51, 52
colour judgement, 63
Columbus, 40
  opinions of American Indians, 41–2
commercialism resulting in savagery, 128
Committee for Nuclear Information, 168
Commodus, Emperor of Rome, 14
Commonwealth Forestry Conference (1962), 154
communications
  with animals, 2
  industrialised and Third World countries, 172
  under imperialism, 127
community development, 3
computers, 160
concentration camps
  Kitchener's for Boers, 126
  Nazi, 140
Congo, 56, 57
  atrocities, 127
  Belgian acquisition, 69
  destruction of village culture, 127
  equatorial forest, 155
  imperialism, 127–8
  pygmies, 167
  slave-trade, 128
  traveller in, 185
Conrad, Joseph, 56, 70, 153, 187, 196
conscience in struggle against savage, 50
conservation, 147
Constitution of United States of America, 82
Cook, Captain James, 105, 118
Cooper, James Fenimore, 88, 132, 191
Coral Island, 129
counter-insurgency, 159
creation myth, 17–18
crime, accusation of whole alien group, 63
Crime and Custom in Savage Society, 196
criminal laws, 100
Cro-Magnon man, 116
Crusades, 21, 178–9
  value of horses, 22
Crystal Palace Exhibition (1851), 110

cultivation and creation of title in land, 163
Cuvier, Georges Léopold, 114–15
cybernetics, 146

da Gama, Vasco, 42, 43
Dahomey
  slavery in, 61
  see also Benin
dance, 142–3
Darling, Sir Frank Fraser, 196
Darwin, Charles, 112, 114, 192
  application of theories to social situation, 134, 135
Darwinism, social, 134, 135, 149
Das Kapital, 112
Davidson, B, 190, 194
de Gobineau, Comte, 110–11
De Golbéry, S M X, 93–4, 190
de Maistre, Comte Joseph, 114
de Tocqueville, Alexis, 163, 164, 197
decadence, 139
Decline and Fall of the Roman Empire, 14
Deerfield Massacre, 87
defence, 3
Defoe, Daniel, 90–2, 194
defoliation of jungles, 157
degeneracy, 116–17
degenerative theory of primitive man, 114, 115, 192
Democracy in America, 163, 197
Denig, Edwin, 85
Descartes, René, 98, 99
Descent of Man, The, 112, 192
Diary from Dixie, A, 186
Discourses on Inequality, 84
Discovery of the Bermudas, Otherwise Called the Isle of Devils, 78, 187
disease
  amongst aborigines, 118
  livestock, 169
  in South Seas, 118
  spread by colonialism, 43
  spread by Europeans, 56
dogs, trained, 44–5
dolphins, 13
Dom João, King of Portugal, 28, 180
domestication of animals, 2, 3
  by Greeks, 12–13
Dominicans, 33
Don Juan, 104
Donner expeditions, 185
Doré, Gustav, 88

INDEX

Dreiser, Theodore, 135, 195
drum culture, 154, 155
Dynasty of Abu, The, 178

Eastern Europe, 147, 171
Ecclesiastical Polity, 82
ecological balance, 160
ecology, 100, 147, 169, 198
economic depression in USA, 149
economics
   American settlers, 122
   Darwin's influence, 134
Ecotopia, 166, 197
Eden, 76
education
   infants, 149
   of primitive men, 132
Egyptians, 9
   deities, 10
Eirik the Red, 37
Eirik's Saga, 37, 39
El Dorado, 76
Eliot, T S, 176
Elizabeth the First, Queen of England, 31
Emerson, Ralph Waldo, 108, 122, 193, 194–5
End of Nature, The, 172
Engels, Friedrich, 111, 192
Enkidu, 5–7
   see also Epic of Gilgamesh
enslavement, 73
environment
   contamination, 168
   poisoning, 171
environmentalism, 151
Epic of Gilgamesh, 5–8, 149, 152, 167, 177, 187
epidemics, 56
Erman, Adolf, 177
Erskine, Lord Thomas, 107, 108
Essay on Civil Government, 82
Essays on the Inequality of the Human Races, 110–11
Ethics, 34
eugenicists, 149
Euripides, 12
Europe: The Emergence of an Idea, 178
Europe
   attitudes towards savagery in Industrial Revolution, 101
   emergence, 190
   expansion, 98–9
   imperialism, 28
   influence of Americas and Africa, 75
   origin of free governments, 83–4
   racial prejudice, 96
   urban poor, 111
Europeans
   accusation of whole alien group with crime, 63
   in Africa, 62–3
   becoming savages by intermarriage, 89
   city dwellers, 137–8
   conflict of culture meeting early Americans, 182
   cult of savage in thought, 101
   explorers' confrontation with savage, 72
   global commerce, 56
   natural goodness of uncivilised man, 64–5
   opinions of Africans, 66–7
   prejudices, 62–4
   relating African tribes with environment, 64
   skin colour prejudice, 62, 63
   soldiers, 125–6
   spread of disease, 56
   superiority feelings over Africa, 93
Ewbank, Thomas, 125
Experiment on a Bird in the Air Pump, 99
exploration
   European confrontation with savage, 72
   Victorian, 65–6
extermination of groups, 148
eye culture, 196

factories
   acid rain emission, 173
   Victorian, 124–5
factory system, 111
Fall of Man, 117
Fanon, Franz, 166
Far West Frontier, The, 185
farmers, 3
   displacement of savage, 163
   on poor land, 167
Fascism, 139–41
fertility ceremonies, 17
feudalism
   European, 23
   in Spain and Portugal, 33
Fitzgerald, C P, 177
Flint, Timothy, 89
food storage, 4
football fans, 162

forest
  clearing, 3
  de Tocqueville's perceptions, 163
  dwellers and attitudes of Spanish to, 30
  influence on people, 153
  replanting, 171
  tribes, 1
  and weather patterns, 154
  see also woods
fossils, 109, 192
Four Stages of Cruelty, The, 100
Frankenstein, 102–4
Franklin, Benjamin, 63–4
Franks, 21–2
Frazer, Sir James, 1, 16–17, 178
freedom, 165–6
French
  intermarriage with Indians in Canada, 89
  Revolution, 101
Freud, Sigmund, 55, 139
frontier
  as moving contract, 133
  in slums, 161
  theory of, 143, 176, 177
  Turner's description, 85
fundamentalists, Biblical, 114
  attack by Darwinian anthropologists, 115
fur traders, 188
Futurism, 140

Gaia, 169
gang rape, 162
Garden of Allah, 76
Gates, Sir Thomas, shipwreck of, 187
Gauguin, Paul, 117, 118, 193
General History of Virginia, New England and the Summer Isles, The, 87
Genesis, 17, 18, 108, 178
genetic determinism, 151
genetic fitness, 150
Genghis Khan, 22
gentes, 116
Géricault, Jean Louis, 73
Germania, 14
Germans, 14–15
  sense of superiority under Hitler, 140
  in state of nature in woods, 83
  tree worship, 17
Gibbon, Edward, 14
Gilgamesh, 5–8, 75
global warming, 171

Godwin, William, 101–2
gold, 28, 33, 44, 46
  discovery in South Africa, 69
Golden Bough, The: A Study in Magic and Religion, 178
Gordon Riots, 101
Gothic writers, 131
Götterdämmerung, 139
governments
  origin of free in northern Europe, 84
  personal terror of scale, 160
  violent techniques of civilised, 160
Graenlendinga Saga, 37–9
Graves, Robert, 179
Great American Desert, 53, 184
Great Plains, The, 184
Great Wall of China, 55, 98
Greeks, 10
  classical ages, 10
  deities, 12
  domestication of animals, 12–13
  philosophers' attitudes towards animals, 13
  population pressure, 177–8
  relationship with natural world, 12
  slavery by, 10–12
  urban society, 10–11
  warrior virtues, 10–11
green
  consciousness, 170, 172
  parties, 160
  pressure groups, 171
greenhouse effect, 160
Greenland, 37, 40
gunpowder empires, 185

Hakluyt, Richard, 47–8, 79, 183
Halieutica, 178
Handlin, Oscar, 137
Hanke, L, 181
Harriot, John, 47
Harriot, Thomas, 183
Harvey, William, 99
Hawkins, John, 181
Hawthorne, Nathaniel, 88–9, 132
Hay, Denys, 178
Headlong Hall, 106–7
Heart of Darkness, 56, 70–2, 187, 196
Hebrew faith, 19
Hemingway, Ernest, 159, 160
Henry the Navigator, Prince of Portugal, 68, 180
herders, 3

INDEX

Herrera, Antonio de, 40, 41, 42, 182, 183
Hindu cosmology, 10
Hiroshima, 168
Hiroshima, Mon Amour, 168
His Natural Life, 120
Hispaniola, 41, 42
historiography, 178
History of Jamaica, 94
Hitler, Adolf, 140, 149
Hobbes, Thomas, 48, 83, 188
hobos, 134
Hofmeyr, Jan H, 148
Hogarth, William, 100
holism, 169
Homer, 10
Homo Ludens, 180
Homo sapiens, 116
Hooker, Richard, 82
horses
  cult of, 22
  in Spanish attacks on Aztecs, 44
Hottentots, 64, 65, 148
Houston, James, 64
Hsia dynasty (China), 8
Huckleberry Finn, 122
Hudson River school of painters, 109
Huizinga, Johan, 29, 180
humans
  bettering of species, 149
  infant education, 149
  parallel behaviour with primate, 150
  sacrifice, 118
  settlement, 1
  sociability, 159
Humboldt, Alexander, 156
Hume, David, 94, 190
hunter
  communication with animals, 2
  mind of, 176
  in nature, 2
hunter-gatherer populations, 1, 166–7
hunting, 31, 170
  mythology, 2
Huxley, Aldous, 143, 151, 164–5, 166

Ibn Khaldun, 21, 22, 179
Ibos, 152
Icelandic sagas, 37–9
ideology of innate difference and superiority, 24
Ik people of Uganda and Kenya, 167
imperialism, 29, 66
  in Africa, 69

Benin, 128
  communications, 127
  dichotomy in, 148
  economic, 69, 147
  foundations for, 67
  necessity to Europeans, 129
  as oppression, 126–9
  political, 147
  taxes, 127
  Victorian, 125–6
In the American Grain, 184
Incas, 46
Incas, The, 182
Indian corn, 48
individual helplessness, 160
industrial civilisation, in America, 53
industrial poisoning, 171, 172
Industrial Revolution, 100
  destruction of village economy, 101
  development of urban masses, 134
industrial system, 111
industrialisation, 124–5
  creation of urban savage, 136
industrialised nations, 147
inflation, 138, 139
inheritance
  favouring of co-operation, 151
  in making of man, 150
Inness, George, 109
Innuit Indians, 136
Instructions, 47, 48, 79, 183
iron workers, 130–1
irrigation, 2
Isabella, Queen of Spain, 35
Islam
  abodes, 20–1
  ideology of innate difference and superiority, 24
  rise of, 15, 20
  slavery in, 23
  trade routes, 23
Island, The, 105

Jamaican rebellion (1760), 61
jazz, 143
Jefferson, Thomas, 84, 109–10, 121, 188
Jemison, Mrs Mary, 87, 162
Jericho, excavations at, 177
Jerusalem, 15, 16
Jesuits, 190
  influence in China, 96–7
Jesus Christ, 15
Jews, 58

INDEX

John of Holywood, 180
Johnson, Edward, 87
Johnson, Samuel, 59, 94
Jung, Carl Gustav, 131–2, 139, 195
Jungle, The, 136, 137, 195

Karlsefni, Thorfinn, 38, 39
Kennedy, Robert, 61
Khmer Rouge, 147
King Philip's War, 87
Kingsley, Mary, 60
Kipling, Rudyard, 145
Kitchener, Horatio Herbert, 126
Koran, 24
Kropotkin, Piotr Alekseyevich, 134
Kubitschek, President of Brazil, 155
Kublai Khan, 55, 96

Lackawanna Valley, The, 109
Lake Regions of Central Africa, The, 186
land possession, 163
landscape cult, 99
language, 26, 172, 180
 in understanding nature of man, 151–2
Las Casas, Bartolomé de, 32, 34, 156, 180–1
Lautréamont (Isidore Ducasse), 117
Lawrence D H, 143–4, 195
laws, relevance of written, 35
Le Vaillant, 64, 65, 186
Leatherstocking novels, 88, 189
Leon-Portilla, Miguel, 182, 183
Leopold, King of the Belgians, 70, 127, 194
leukaemia, 168
Lévi-Strauss, C, 26, 152, 156
Leviathan, 83
Leviathan, 48
Lewis, Oscar, 161
Lewis, R W B, 191
Liao-shih, 9
Life in Ancient Egypt, 177
Life on the Mississippi, 122, 193
Lima, 158
Linnaeus, Carolus, 94
livestock
 abuse, 124
 degeneration in America, 84
 diseases, 169
 treatment, 178
Livingstone, David, 65, 69, 194
Locke, John, 82, 193

London, Jack, 111, 135–6, 158, 164, 192, 195
London mob, 111
Long, Edward, 94
Lord of Misrule, 51
Lorenz, Konrad, 149–50
Louisiana Purchase, 121
Lovelock, James, 69, 197
Lubbock, Sir John, 115
Luddites, 101
Lusiads, The, 42
Lytton, Bulwer, 110

machines, 123, 124
Mailer, Norman, 160
Malayan rubber trade, 155
Malinowski, Bronislaw, 113, 141, 145, 195
Malraux, André, 142, 195
Manchild in the Promised Land, 161
Man's Responsibility for Nature, 178
marginal land, 166
Marinetti, Filippo Tommaso, 140
Marx, Karl, 111, 123, 192
 Darwin's influence, 134
 influence on Vietnam, 146
Marx, Leo, 190
Matabele, 148
matriarchy, 142
Mbuti pygmies, 167
mechanistic view of world, 98
Mediterranean civilisation, 9
 rise of Islam, 20
 rise of world religions, 15
Mediterranean and the Mediterranean World in the Age of Philip II, 178
Melincourt, 107
Mesopotamia, 4–5
metals, use in weapons, 44
metropolis, 159
Mexican Indians, 182
Mexico, Utopian communities, 77
Midgley, Mary, 190
military drill, 125
military obedience, 139
military power, Victorian, 125
Ming dynasty (China), 56
missionaries
 in Africa, 95
 in Polynesia, 118
 treatment of Indians, 156
 views of behaviour of pastoral societies, 128–9
Mister Johnson, 145, 196

mob assault, 159
Mobutu, Sese Seko, President of Zaïre, 146–7
Mongols, 22, 150
   conquest of China, 55
monists, 149
monogamy, European, 61
Montaigne, Michel Eyquem de, 98, 180, 190
Montesquieu, Charles Louis, 83
More, Sir Thomas, 76, 78, 180, 187
Morgan, Lewis Henry, 115
Morris, Desmond, 150, 197
Morton, 51–2
Motecuhzoma, Aztec emperor, 45, 46
mountain men, 53–4, 184–5, 189–90
mugging, 162
Muhammad, prophet, 20
music, black, 143
Muslims
   Crusades, 21
   see also Islam
mythology of hunting societies, 2

Nagasaki, 168
Naked Ape, The, 150, 169, 197
natural history
   interest in nineteenth century America, 109
   mutual aid in survival of species, 134
Natural History of Selborne, The, 100
natural sciences, 191
natural selection theory, 112
natural world, separation of humans from in Industrial Revolution, 101
nature
   authority of man over, 121
   bonding with, ix
   man debasing relationship with, 169
   of men in relation to civilisation and wilderness, 149
Nazi
   concentration camps, 58
   party, 140
Nevermore, 118
New Guinea, 196
New World
   interaction with Old, 78
   Shakespeare's exploration of myth, 79
New York, violence in, 162
Newton, Isaac, 99
Nigeria, 152

Nile
   civilisations, 9
   source, 65–6
Noah, 18–19
Noble Savage
   Africans as, 94, 95
   of South Seas, 118
   and Utopia, 77
Norris, Frank, 123
Norse expeditions, 181–2
Norsemen, 37
North America
   displacement of Indians, 74
   race relations, 74
   see also America; United States of America
Nova Atlantis, 78
novelists, 194
novels
   American realist school, 135
   Australian life, 120
   inspired by Africa, 95
   inspired by American Indians, 88–9
nuclear
   power, 168
   weapons, 160

obsidian, 44
Octopus, The, 123
Of Plymouth Plantation, 184
Of Taste, 99
Ojibway Indians (North America), 25
Old Calabar, 60, 186
On Aggression, 149
order imposed on nature and savage, 121
Oregon Trail, The, 109, 131
Origin of Civilisation, The, 115
Ossian, 13, 178
ostracism, 36
Othello, 81–2
ozone layer, 158, 171

painters *see* artists
painting, rock, 1
palaeontology, 109, 192
paradise, tangible, 76
parasites, 1, 176
Paris World Fair (1889), 117
Park, Mungo, 186
Parkman, Francis, 131, 132
parks, 99
Passmore, John, 178
pastoral ideal, 190

INDEX

pastoral philosophers, 122
patriotism, 139
patronage, 177
Paul III, Pope, 33
Peacock, Thomas Love, 106, 107
penal code, 100, 191
penal colonies, 119–20
People of the Abyss, The, 135, 192, 195
People of Plenty, 124
Percy, George, 183
perfect place, search for, 76
Peterloo, 191
petrol engine, 123
pets, 170
Philosophy of Manufacturers, The, 124
photosynthesis, 154
Pilgrims, 49–51
Pilgrims, 93
pioneer trappers, 53–4
Pioneers, The, 88, 108
plantation owners, 61
  erosion of moral standards, 61–2
plantations, 157
Plato, 11, 13, 55, 185
  creation of Atlantis, 75
play, human, 31
Pliny, 178
ploughing, 2–3
Plumed Serpent, The, 143
Plutarch, 14
Plymouth plantation, 49, 50, 51, 184
Pocahontas, 51, 52
Poe, Edgar Allen, 110
poetry, 104–5
  mythical, 152
police, secret, 162
political liberty, 165–6
Politics, 34, 178
pollution, 147, 158, 160, 168, 171, 172, 173
Polo, Marco, 96
polygamy, African, 61
Polynesia, 117, 118
Pope, Alexander, 99, 190
Pope-Hennessy, J, 186
population explosion, 169
Portilla *see* Leon-Portilla, Miguel
Portuguese
  slavery by, 32, 56
  spread of disease, 56
  voyages, 28–9, 30
potatoes, 49
Prairie, The, 88

prairies, 123, 184
precivilised people, 116
predators, 176
prehistoric Europe, 4
prehistoric species, 116
preliterate peoples, survival, 166
Prester John, 95
prey, dependence on, 2
Primitive Culture, 113, 145, 176
Primitive Man as Philosopher, 142, 179
primitive people
  degenerative theory, 114, 115, 192
  dispossession by civilisation, 164
  education, 132
  nostalgia for, 131
  observation by urban westerners, 43
  realities, 25
  society, 142
Primitive Art, 142
Promised Land, 109
property
  ownership, 115, 163
  value in labour, 82
Prospero, 78–9, 80, 81
Protestant Reformation, 172
Protestant slavers, 58
psychiatry, 139
  West Africa, 153
Purchas, Samuel, 93
Puritans, 50–1, 184, 189
  examination of ambiguities of conscience, 89
  fear of mountain men, 53–4
  move to plains, 52, 53
  threat of savages, 52
  view of Indians as agent of Satan, 51
pygmies, 64, 167
  of Zaire, 146–7
Pythagoras, 13

Quiroga, Vasco de, 77

race, nineteenth century philosophy, 110
racial prejudice in Europe, 62, 63, 96
racist thought, 192
Radcliffe, Mrs Ann, 131
radiation, 168
Radin, Paul, 24–5, 142, 179–80
Raft of the Medusa, 73
ragtime, 143
railways, 122, 123, 193
  and travel, 130
reason, human, 99

INDEX

records, Mesopotamia, 4
Regulations for Field Forces in South Africa, 125
Republic, 55, 75, 185
Requerimiento, 35
rhythm, 142–3
Ricci, Matteo, 96–7
Richard of Devizes, 21
Riot Acts, 191
ritual
  beheading, 186
  centre construction, 4
roads, 123–4
Robinson Crusoe, 90–2, 190
Rocky Mountains, 53–4
Rolfe, John, 51, 52
Roman civilisation
  barbarity, 13
  circuses, 13
  games, 14
  plays, 14
  sensibility towards animals, 14
Romantics, 191
  revolution against industrialism, 101
  view of animals' relationship with humans, 107
  view of primitive being, 106
Rome, sack by Spanish troops, 33, 34
Rousseau, Jean Jacques, 84, 85, 94, 167, 197
rubber
  production, 127
  tapping in Amazon, 155
Ruskin, John, 73
Russia, 147, 171
  population pressure, 159
Russian steppes, 124

sacrifice, substitutes for, 17
St Augustine, 16, 19
St Thomas Aquinas, 23
Saladin, 21
Samuel, Herbert, 127
Sanderson, Ivan, 178
Saukee Indians, 53
savage
  classification by Tylor, 113
  cult of, 104, 105, 140
  degenerative theory, 114, 115, 192
  derivation of word, 174–5
  language in modern world, 172
  perception of civilised people, 167
  survival in modern world, 166
  theory of, 113
Savage Society, 141
Savage, The, 175
savageness, 159, 160
savagery
  amongst men in groups, 63
  see also violence in cities
Scarlet Letter, The, 89
Schell, Adam, 97
scientific instruments, American Indians' perception, 47
scientific method, 99
self-sacrifice, 139
senses, world of, 25
Sepulvéda, Juan Ginés de, 32, 33
serfs, 23
  in Iberia, 33
Service for the Protection of Indians, 155
Shaftesbury, Earl of, 191
Shakespeare, William, 36, 78, 79, 179, 188
  use of dark savage, 81–2
shanty-towns, 158
Shao Yung, 27
Shelley, Mary, 101, 102
Shelley, Percy Bysshe, 101, 102
Shintoism, 2
Sibley, Henry Hastings, 85
Silent Spring, The, 154, 168, 197
Silk Road, 55
simplicity, primitive, 117
Sinclair, Upton, 136, 137, 164, 195
Sins of the Fathers, 186
Sister Carrie, 135, 195
skin colour prejudice, 62, 63
Skraelings, 37, 38
slash-and-burn agriculture, 1, 48
Slave Ship, The, 73
slave-trade, 181
  apologists for, 59–60
  Benin, 128
  Congo, 128
  end, 65, 68
  guilt, 73
  justifications, 59, 60
slavery, 3, 5, 185, 186
  of Africans, 58
  Aristotle's writings on, 34
  by Egyptians, 9
  by Greeks, 10, 11
  by Portuguese, 32, 157
  by Spaniards, 29, 32, 42
  by Sumerians, 5

210

INDEX

Chinese expeditions, 179
commercialism, 58, 59
  of Congolese, 56
  corruption in, 58
  economics, 60, 157
  effects on novelists, 95
  erosion of moral standards, 61–2
  in Islam, 23
  persistence in America, 110
  political economy, 61
slaves
  in plantation house, 61
  savagery to, 62
  uprisings, 61
Slaves Throwing Overboard the Dead and Dying – Typhoon Coming On, 73
slum populations, 134, 136
slums, 135–6, 137, 164
  frontier in, 161
smallpox, 43, 56
Smith, Adam, 97
Smith, Captain John, 87, 189
Smith, Henry Nash, 176, 195
Snelgrave, William, 186
Social Contract, The, 150
social legislation, 149
socialism, 136, 149
society
  divisions, 142
  of primitive man, 142
  progress of, 15
  reality to primitive man, 25
Society for the Prevention of Cruelty to Animals, 108
Sociobiology, 150–1
Socrates, 13
Solomon, Job ben, 59, 96, 190
Song of Myself, 132–3
souls
  of animate beings, 17
  equality in Christian doctrine, 23
sounds, relevance to forest people, 153
South Africa, 68
  apartheid, 147
  discovery of gold, 69
  displacement of blacks by colonists, 74
  European soldiers in, 125–6
  National Game Parks, 147
  native policy, 148
  race relations, 74
South Seas, 105, 117, 118
Spain
  atrocities to American Indians, 34–5
  attitudes to American Indians, 33–4
  attitudes to forest dwellers, 30
  colonialism, 34–5
  conquistadors, 34, 46
  debate on American Indians, 29, 32
  early opinions of America, 40–1
  gold in New World expeditions, 46
  ineffectiveness of laws applying to colonies, 34–5
  medieval, 29–30
  Mexican Indians' reaction to, 182
  rise of power, 29
  slavery by, 32, 42
  spreading of Catholic faith to Americas, 32, 33, 90
  voyages to Africa and America, 30
species
  degeneration, 116
  Linnaeus' classification, 94
  mutual aid in survival, 134
  relationships, 178
Spencer, Herbert, 134
Spengler, Oswald, 157
Sphaera Mundi, 180
Stanley, Henry Morton, 125, 128, 194
stateless people, 138
Stiens (Cambodia), 2
Stowe, Harriet Beecher, 110
Strabo, 13
Studies in Classic American Literature, 195
Sublimus Deus, 33
Sudanese slaves, 9
Sumerian temples, 4
Survival communities, 166
Swift, Jonathan, 92
Syntax Structure, 180
syphilis, 43
Systema Natura, 94

taboos, destruction of aborigine, 119
Tacitus, 1, 78, 83
  myth of noble savage, 14
Tahiti, 117, 118
T'ang Empire (China), 22
Taoism, 2
taxes, 4
Taylor, Thomas, 107
technology, 122, 123–4, 190, 192
  revolution, 160
  Victorian imperialism, 125
Tempest, The, 78–81, 179, 188
Tenochtitlan, 45
Territorial Imperative, The, 150

INDEX

theocracy, 4
Theodore, Emperor of Ethiopia, 69, 159
Third Reich, 140
Third World, 147
   cities, 157–8
   overpopulation of cities, 196
   response to green arguments, 172
Thomas, Dylan, ix
Thoreau, Henry David, 122, 123, 131, 132, 193
thought police, 164
timber cutting and clearing, 3
Timon of Athens, 36
Titus Andronicus, 82
tools, 3
Topsoil and Civilisation, 170–1, 198
torture, 159, 162, 197
totalitarianism, 165
tourism, 129, 130, 147, 194
   escape from city life, 163
tractors, 124
Trajan, Emperor of Rome, 14
tramps, 134, 137
Transamazon Highway, 155
Transcendentalists of Boston, 131, 132
trapping, 53–4
   by American Indians, 86
travellers, 130, 194
travellers' tales, 76
tree-worship, 1, 17, 152
trees
   Buddhist attitudes, 23
   in leisure industry, 162
   link between men and, 16–17
   mass consciousness of value, 154
   sacred, 152
   use for defence, 35–6
tribal organisation, 4
tribalism, modern, 166
Trobriand Islanders, 113, 141–2, 195
   matriarchy, 142
   rule, 141
True Relation, 87
Turnbull, Colin M, 197
Turner, E S, 191
Turner, Frederick Jackson, 85, 133, 143, 176, 195
Turner, J M W, 73
Twain, Mark, 122, 193
Tylor, Sir Edward Burnett, 113, 115, 145, 176, 192
Types of Mankind, 110
Tyson, Edward, 64

underdeveloped countries, 147
unemployment, 138, 139
United States of America
   constitution, 82
   organisation, 121
   social legislation, 149
Updike, John, 168, 197
urban civilisation, lost innocence, 76
urban ghettoes, 149
urban man, 27, 131
urban poor, 111
urban savage, 136
urban unrest, 101
Ure, Andrew, 124
Uruk, 75
Utnapishtim, 75
Utopia, 76, 180, 187
Utopia, 117, 164, 165
   and Noble Savage, 77
   Renaissance, 78
Utopianism, 76

Valladolid, 29, 32, 180
Vanishing Primitive Man, 197
Verbeist, Ferdinand, 97
Vespucci, Amerigo, 78, 182
Victorian age
   belief in superiority, 168
   decadence, 117
   imperialism, 125–6
   industrial development, 124–5
   judgement by pioneers, 129
   missionaries, 128
   virtue in product of the woods, 132
Vietnam, 126, 157
   Marxist influence, 146
Vikings, 37
village cultures, 3
Vindication of the Rights of Brutes, 107
Vinland, 181–2
   discovery of, 37
   Norse journeys to, 38
violence in cities, 162, 163, 164
Virgin Land: The American West as Symbol and Myth, 176
Virginia (America), 47, 78
vivisection, 99, 169
Voices of Silence, The, 142, 195
Voltaire, François Marie, 94, 97
voodoo, 61

walls, defensive, 177
   see also Great Wall of China

INDEX

Walpole, Horace, 130, 194
Waning of the Middle Ages, The, 29, 180
war
  civilised man against savages, 121
  religious justification, 45
  state in primitive society, 83
  technology, 47
Ward, Nathaniel, 50
Washington, George, 132
weapons, 44
weather patterns, 154
Webb, Walter P, 184
Weiner, J S, 197
West Africa
  British presence, 68
  relevance of sound to forest dwellers, 153
Western society, recapture of savages, 160
wheat, 123
Wheeler, Charles, 95
White, Gilbert, 100, 190
White Goddess, The, 179
White, John, 47
white man's burden, 145
whites, effects of industrial system of division from blacks, 111
Whitman, Walt, 132-3
wilderness, 131
  American, 133
  American perceptions, 191
  hunting ground as, 163
  London's equation with ghetto, 135
  nature of men in relation to, 149
  preservation, 147
  T S Eliot's *Four Quartets*, 176
  urban, 134
wilding, 162
wildlife films, 170
Williams, Tennessee, 173

Williams, William Carlos, 184
Wilsher, Peter, 196
Wilson, Edward O, 150, 151, 196
Winter's Tale, A, 31
witch-doctors, 61
Wollstonecraft, Mary, 107-8
Wolseley, Garnet, 125
Wonder-Working Providence, 87
woods
  origins from perceived by Victorians, 132
  regrowth round cities, 162
  virtues, 134
  see also forest
words
  written, 24-6
  see also language
Wordsworth, William, 132, 191
working class
  exploitation, 111
  social inferiority, 112
World War I, 126, 138
  aftermath, 138
  questioning of motives, 139
  regression to barbaric values, 139
World War II, 145
World Wide Fund for Nature, 198
World a Workshop, The, 125
Wright, Joseph, 99
writing, 24-6, 179

Yggdrasil, 163, 171
Ying-wei-chih, 9
Yoruba, 152

Zaire, 146-7
Zulu Wars, 194
Zulus, 148
  impis, 125-6

Also by Andrew Sinclair

FICTION

The Breaking of Bumbo
My Friend Judas
The Project
The Hallelujah Bum
The Raker
A Patriot for Hire
The Facts in the Case of E. A. Poe
Beau Bumbo
Gog
Magog
King Ludd

NONFICTION

Prohibition: the Era of Excess
The Better Half: the Emancipation of the American Woman
The Available Man: Warren Gamaliel Harding
The Concise History of the United States
Che Guevara
The Last of the Best
The Savage
Dylan: Poet of His People
Jack London
John Ford
Corsair: the Life of J. P. Morgan
The Other Victoria
Sir Walter Raleigh and the Age of Discovery
The Red and the Blue
War Like a Wasp: the Lost Decade of the Forties
The Need to Give

DRAMA

Adventures in the Skin Trade (play)
Under Milk Wood (screenplay)

TRANSLATION

The Greek Anthology

ANTHOLOGY

The War Decade

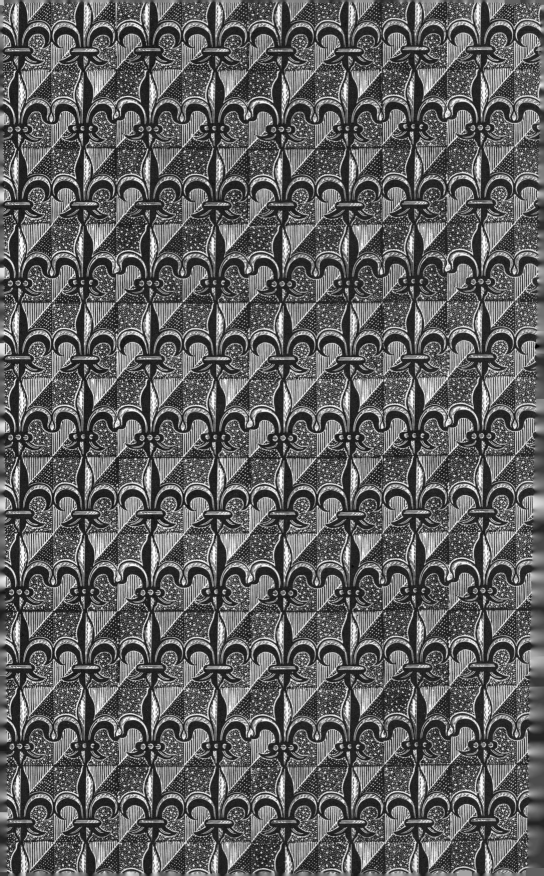